THE STORY OF EVOLUTION

By

Joseph McCabe

1912

PREFACE

An ingenious student of science once entertained his generation with a theory of how one might behold again all the stirring chapters that make up the story of the earth. The living scene of our time is lit by the light of the sun, and for every few rays that enter the human eye, and convey the image of it to the human mind, great floods of the reflected light pour out, swiftly and indefinitely, into space. Imagine, then, a man moving out into space more rapidly than light, his face turned toward the earth. Flashing through the void at, let us say, a million miles a second, he would (if we can overlook the dispersion of the rays of light) overtake in succession the light that fell on the French Revolution, the Reformation, the Norman Conquest, and the faces of the ancient empires. He would read, in reverse order, the living history of man and whatever lay before the coming of man.

Few thought, as they smiled over this fairy tale of science, that some such panoramic survey of the story of the earth, and even of the heavens, might one day be made in a leisure hour by ordinary mortals; that in the soil on which they trod were surer records of the past than in its doubtful literary remains, and in the deeper rocks were records that dimly lit a vast abyss of time of which they never dreamed. It is the supreme achievement of modern science to have discovered and deciphered these records. The picture of the past which they afford is, on the whole, an outline sketch. Here and there the details, the colour, the light and shade, may be added; but the greater part of the canvas is left to the more skilful hand of a future generation, and even the broad lines are at times uncertain. Yet each age would know how far its scientific men have advanced in constructing that picture of the growth of the heavens and the earth, and the aim of the present volume is to give, in clear and plain language, as full an account of the story as the present condition of our knowledge and the limits of the volume will allow. The author has been for many years interested in the evolution of things, or the way in which suns and atoms, fishes and flowers, hills and elephants, even man and his institutions, came to be what they are. Lecturing and writing on one or other phase of the subject have, moreover, taught him a language which the inexpert seem to understand, although he is not content merely to give a superficial description of the past inhabitants of the earth.

The particular features which, it is hoped, may give the book a distinctive place in the large literature of evolution are, first, that it includes the many evolutionary discoveries of the last few years, gathers its material from the score of sciences which confine themselves to separate aspects of the universe, and blends all these facts and discoveries in a more or less continuous chronicle of the life of the heavens and the earth. Then the author has endeavoured to show, not merely how, but why, scene succeeds scene in the chronicle of the earth, and life slowly climbs from level to level. He has taken nature in the past as we find it to-day: an interconnected whole, in which the changes of land and sea, of heat and cold, of swamp and hill, are faithfully reflected in the forms of its living population. And, finally, he has written for those who are not students of science, or whose knowledge may be confined to one branch of science, and used a plain speech which assumes no previous knowledge on the reader's part.

For the rest, it will be found that no strained effort is made to trace pedigrees of animals and plants when the material is scanty; that, if on account of some especial interest disputable or conjectural speculations are admitted, they are frankly described as such; and that the more important differences of opinion which actually divide astronomers, geologists, biologists, and anthropologists are carefully taken into account and briefly explained. A few English and American works are recommended for the convenience of those who would study particular chapters more closely, but it has seemed useless, in such a work, to give a bibliography of the hundreds of English, American, French, German, and Italian works which have been consulted.

CONTENTS

THE STORY OF EVOLUTION

CHAPTER I. THE DISCOVERY OF THE UNIVERSE

The beginning of the victorious career of modern science was very largely due to the making of two stimulating discoveries at the close of the Middle Ages. One was the discovery of the earth: the other the discovery of the universe. Men were confined, like molluscs in their shells, by a belief that they occupied the centre of a comparatively small disk—some ventured to say a globe—which was poised in a mysterious way in the middle of a small system of heavenly bodies. The general feeling was that these heavenly bodies were lamps hung on a not too remote ceiling for the purpose of lighting their ways. Then certain enterprising sailors—Vasco da Gama, Maghalaes, Columbus—brought home the news that the known world was only one side of an enormous globe, and that there were vast lands and great peoples thousands of miles across the ocean. The minds of men in Europe had hardly strained their shells sufficiently to embrace this larger earth when the second discovery was reported. The roof of the world, with its useful

1

little system of heavenly bodies, began to crack and disclose a profound and mysterious universe surrounding them on every side. One cannot understand the solidity of the modern doctrine of the formation of the heavens and the earth until one appreciates this revolution.

Before the law of gravitation had been discovered it was almost impossible to regard the universe as other than a small and compact system. We shall see that a few daring minds pierced the veil, and peered out wonderingly into the real universe beyond, but for the great mass of men it was quite impossible. To them the modern idea of a universe consisting of hundreds of millions of bodies, each weighing billions of tons, strewn over billions of miles of space, would have seemed the dream of a child or a savage. Material bodies were "heavy," and would "fall down" if they were not supported. The universe, they said, was a sensible scientific structure; things were supported in their respective places. A great dome, of some unknown but compact material, spanned the earth, and sustained the heavenly bodies. It might rest on the distant mountains, or be borne on the shoulders of an Atlas; or the whole cosmic scheme might be laid on the back of a gigantic elephant, and—if you pressed—the elephant might stand on the hard shell of a tortoise. But you were not encouraged to press.

The idea of the vault had come from Babylon, the first home of science. No furnaces thickened that clear atmosphere, and the heavy-robed priests at the summit of each of the seven-staged temples were astronomers. Night by night for thousands of years they watched the stars and planets tracing their undeviating paths across the sky. To explain their movements the priest-astronomers invented the solid firmament. Beyond the known land, encircling it, was the sea, and beyond the sea was a range of high mountains, forming another girdle round the earth. On these mountains the dome of the heavens rested, much as the dome of St. Paul's rests on its lofty masonry. The sun travelled across its under-surface by day, and went back to the east during the night through a tunnel in the lower portion of the vault. To the common folk the priests explained that this framework of the world was the body of an ancient and disreputable goddess. The god of light had slit her in two, "as you do a dried fish," they said, and made the plain of the earth with one half and the blue arch of the heavens with the other.

So Chaldaea lived out its 5000 years without discovering the universe. Egypt adopted the idea from more scientific Babylon. Amongst the fragments of its civilisation we find representations of the firmament as a goddess, arching over the earth on her hands and feet, condemned to that eternal posture by some victorious god. The idea spread amongst the smaller nations which were lit by the civilisation of Babylon and Egypt. Some blended it with coarse old legends; some, like the Persians and Hebrews, refined it. The Persians made fire a purer and lighter spirit, so that the stars would need no support. But everywhere the blue vault hemmed in the world and the ideas of men. It was so close, some said, that the birds could reach it. At last the genius of Greece brooded over the whole chaos of cosmical speculations.

The native tradition of Greece was a little more helpful than the Babylonian teaching. First was chaos; then the heavier matter sank to the bottom, forming the disk of the earth, with the ocean poured round it, and the less coarse matter floated as an atmosphere above it, and the still finer matter formed an "aether" above the atmosphere. A remarkably good guess, in its very broad outline; but the solid firmament still arched the earth, and the stars were little undying fires in the vault. The earth itself was small and flat. It stretched (on the modern map) from about Gibraltar to the Caspian, and from Central Germany—where the entrance to the lower world was located—to the Atlas mountains. But all the varied and conflicting culture of the older empires was now passing into Greece, lighting up in succession the civilisations of Asia Minor, the Greek islands, and then Athens and its sister states. Men began to think.

The first genius to have a glimpse of the truth seems to have been the grave and mystical Pythagorus (born about 582 B.C.). He taught his little school that the earth was a globe, not a disk, and that it turned on its axis in twenty-four hours. The earth and the other planets were revolving round the central fire of the system; but the sun was a reflection of this central fire, not the fire itself. Even Pythagoras, moreover, made the heavens a solid sphere revolving, with its stars, round the central fire; and the truth he discovered was mingled with so much mysticism, and confined to so small and retired a school, that it was quickly lost again. In the next generation Anaxagoras taught that the sun was a vast globe of white-hot iron, and that the stars were material bodies made white-hot by friction with the ether. A generation later the famous Democritus came nearer than any to the truth. The universe was composed of an infinite number of indestructible particles, called "atoms," which had gradually settled from a state of chaotic confusion to their present orderly arrangement in large masses. The sun was a body of enormous size, and the points of light in the Milky Way were similar suns at a tremendous distance from the earth. Our universe, moreover, was only one of an infinite number of universes, and an eternal cycle of destruction and re-formation was running through these myriads of worlds.

By sheer speculation Greece was well on the way of discovery. Then the mists of philosophy fell between the mind of Greece and nature, and the notions of Democritus were rejected with disdain; and then, very speedily, the decay of the brilliant nation put an end to its feverish search for truth. Greek culture passed to Alexandria, where it met the remains of the culture of Egypt, Babylonia, and Persia, and one more remarkable effort was made to penetrate the outlying universe before the night of the Middle Ages fell on the old world.

Astronomy was ardently studied at Alexandria, and was fortunately combined with an assiduous study of mathematics. Aristarchus (about 320-250 B.C.) calculated that the sun was 84,000,000 miles away; a vast expansion of the solar system and, for the time, a remarkable approach to the real figure (92,000,000) Eratosthenes (276-196 B.C.) made an extremely good calculation of the size of the earth, though he held it to be the centre of a small universe. He concluded that it was a globe measuring 27,000 (instead of 23,700) miles in circumference. Posidonius (135-51 B.C.) came even nearer with a calculation that the circumference was between 25,000 and 19,000 miles; and he made a fairly correct estimate of the diameter, and therefore distance, of the sun. Hipparchus (190-120 B.C.) made an extremely good calculation of the distance of the moon.

By the brilliant work of the Alexandrian astronomers the old world seemed to be approaching the discovery of the universe. Men were beginning to think in millions, to gaze boldly into deep abysses of space, to talk of vast fiery globes that made the earth insignificant But the splendid energy gradually failed, and the long line was closed by Ptolemaeus, who once more put the earth in the centre of the system, and so imposed what is called the Ptolemaic system on Europe. The keen school-life of Alexandria still ran on, and there might have been a return to the saner early doctrines, but at last Alexandrian culture was extinguished in the blood of the aged Hypatia, and the night fell. Rome had had no genius for science; though Lucretius gave an immortal expression to the views of Democritus and Epicurus, and such writers as Cicero and Pliny did great service to a later age in preserving fragments of the older discoveries. The curtains were once more drawn about the earth. The glimpses which adventurous Greeks had obtained of the great outlying universe were forgotten for a thousand years. The earth became again the little platform in the centre of a little world, on which men and women played their little parts, preening themselves on their superiority to their pagan ancestors.

I do not propose to tell the familiar story of the revival at any length. As far as the present subject is concerned, it was literally a Renascence, or re-birth, of Greek ideas. Constantinople having been taken by the Turks (1453), hundreds of Greek scholars, with their old literature, sought refuge in Europe, and the vigorous brain of the young nations brooded over the ancient speculations, just as the vigorous young brain of Greece had done two thousand years before. Copernicus (1473-1543) acknowledges that he found the secret of the movements of the heavenly bodies in the speculations of the old Greek thinkers. Galilei (1564-1642) enlarged the Copernican system with the aid of the telescope; and the telescope was an outcome of the new study of optics which had been inspired in Roger Bacon and other medieval scholars by the optical works, directly founded on the Greek, of the Spanish Moors. Giordano Bruno still further enlarged the system; he pictured the universe boldly as an infinite ocean of liquid ether, in which the stars, with retinues of inhabited planets, floated majestically. Bruno was burned at the stake (1600); but the curtains that had so long been drawn about the earth were now torn aside for ever, and men looked inquiringly into the unfathomable depths beyond. Descartes (1596-1650) revived the old Greek idea of a gradual evolution of the heavens and the earth from a primitive chaos of particles, taught that the stars stood out at unimaginable distances in the ocean of ether, and imagined the ether as stirring in gigantic whirlpools, which bore cosmic bodies in their orbits as the eddy in the river causes the cork to revolve.

These stimulating conjectures made a deep impression on the new age. A series of great astronomers had meantime been patiently and scientifically laying the foundations of our knowledge. Kepler (1571-1630) formulated the laws of the movement of the planets; Newton (1642-1727) crowned the earlier work with his discovery of the real agency that sustains cosmic bodies in their relative positions. The primitive notion of a material frame and the confining dome of the ancients were abandoned. We know now that a framework of the most massive steel would be too frail to hold together even the moon and the earth. It would be rent by the strain. The action of gravitation is the all-sustaining power. Once introduce that idea, and the great ocean of ether might stretch illimitably on every side, and the vastest bodies might be scattered over it and traverse it in stupendous paths. Thus it came about that, as the little optic tube of Galilei slowly developed into the giant telescope of Herschel, and then into the powerful refracting telescopes of the United States of our time; as the new science of photography provided observers with a new eye—a sensitive plate that will register messages, which the human eye cannot detect, from far-off regions; and as a new instrument, the spectroscope, endowed astronomers with a power of perceiving fresh aspects of the inhabitants of space, the horizon rolled backward, and the mind contemplated a universe of colossal extent and power.

Let us try to conceive this universe before we study its evolution. I do not adopt any of the numerous devices that have been invented for the purpose of impressing on the imagination the large figures we must use. One may doubt if any of them are effective, and they are at least familiar. Our solar system—the family of sun and planets which had been sheltered under a mighty dome resting on the hill-tops—has turned out to occupy a span of space some 16,000,000,000 miles in diameter. That is a very small area in the new universe. Draw a circle, 100 billion miles in diameter, round the sun, and you will find that it contains only three stars besides the sun. In other words, a sphere of space measuring 300 billion miles in circumference—we will not venture upon the number of cubic miles—contains only four stars (the sun, alpha Centauri, 21,185 Lalande, and 61 Cygni). However, this part of space seems to be below the average in point of population, and we must adopt a different way of estimating the magnitude of the universe from the number of its stellar citizens.

Beyond the vast sphere of comparatively empty space immediately surrounding our sun lies the stellar universe into which our great telescopes are steadily penetrating. Recent astronomers give various calculations, ranging from 200,000,000 to 2,000,000,000, of the number of stars that have yet come within our faintest knowledge. Let us accept the modest provisional estimate of 500,000,000. Now, if we had reason to think that these stars were of much the same size and brilliance as our sun, we should be able roughly to calculate their distance from their faintness. We cannot do this, as they differ considerably in size and intrinsic brilliance. Sirius is more than twice the size of our sun and gives out twenty times as much light. Canopus emits 20,000 times as much light as the sun, but we cannot say, in this case, how much larger it is than the sun. Arcturus, however, belongs to the same class of stars as our sun, and astronomers conclude that it must be thousands of times larger than the sun. A few stars are known to be smaller than the sun. Some are, intrinsically, far more brilliant; some far less brilliant.

Another method has been adopted, though this also must be regarded with great reserve. The distance of the nearer stars can be positively measured, and this has been done in a large number of cases. The proportion of such cases to the whole is still very small, but, as far as the results go, we find that stars of the first magnitude are, on the average, nearly 200 billion miles away; stars of the second magnitude nearly 300 billion; and stars of the third magnitude 450 billion. If this fifty per cent increase of distance for each lower magnitude of stars were certain and constant, the stars of the eighth magnitude would be 3000 billion miles away, and stars of the sixteenth magnitude would be 100,000 billion miles away; and there are still two fainter classes of stars which are registered on long-exposure photographs. The mere vastness of these figures is immaterial to the astronomer, but he warns us that the method is uncertain. We may be content to conclude that the starry universe over which our great telescopes keep watch stretches for thousands, and probably tens of thousands, of billions of miles. There are myriads of stars so remote that, though each is a vast incandescent globe at a temperature of many thousand degrees, and though their light is concentrated on the mirrors or in the lenses of our largest telescopes and directed upon the photographic plate at the rate of more than 800 billion waves a second, they take several hours to register the faintest point of light on the plate.

When we reflect that the universe has grown with the growth of our telescopes and the application of photography we wonder whether we may as yet see only a fraction of the real universe, as small in comparison with the whole as the Babylonian system was in comparison with ours. We must be content to wonder. Some affirm that the universe is infinite; others that it is limited. We have no firm ground in science for either assertion. Those who claim that the system is limited point out that, as the stars decrease in brightness, they increase so enormously in number that the greater faintness is more than compensated, and therefore, if there were an infinite series of magnitudes, the midnight sky would be a blaze of light. But this theoretical reasoning does not allow for dense regions of space that may obstruct the light, or vast regions of vacancy between vast systems of stars. Even apart from the evidence that dark nebulae or other special light-absorbing regions do exist, the question is under discussion in science at the present moment whether light is not absorbed in the passage through ordinary space. There is reason to think that it is. Let us leave precarious speculations about finiteness and infinity to philosophers, and take the universe as we know it.

Picture, then, on the more moderate estimate, these 500,000,000 suns scattered over tens of thousands of billions of miles. Whether they form one stupendous system, and what its structure may be, is too obscure a subject to be discussed here. Imagine yourself standing at a point from which you can survey the whole system and see into the depths and details of it. At one point is a single star (like our sun), billions of miles from its nearest neighbour, wearing out its solitary life in a portentous discharge of energy. Commonly the stars are in pairs, turning round a common centre in periods that may occupy hundreds of days or hundreds of years. Here and there they are gathered into clusters, sometimes to the number of thousands in a cluster, travelling together over the desert of space, or trailing in lines like luminous caravans. All are rushing headlong at inconceivable speeds. Few are known to be so sluggish as to run, like our sun, at only 8000 miles an hour. One of the "fixed" stars of the ancients, the mighty Arcturus, darts along at a rate of more than 250 miles a second. As they rush, their surfaces glowing at a temperature anywhere between 1000 and 20,000 degrees C., they shake the environing space with electric waves from every tiny particle of their body at a rate of from 400 billion to 800 billion waves a second. And somewhere round the fringe of one of the smaller suns there is a little globe, more than a million times smaller than the solitary star it attends, lost in the blaze of its light, on which human beings find a home during a short and late chapter of its history.

Look at it again from another aspect. Every colour of the rainbow is found in the stars. Emerald, azure, ruby, gold, lilac, topaz, fawn—they shine with wonderful and mysterious beauty. But, whether these more delicate shades be really in the stars or no, three colours are certainly found in them. The stars sink from bluish white to yellow, and on to deep red. The immortal fires of the Greeks are dying. Piercing the depths with a dull red glow, here and there, are the dying suns; and if you look closely you will see, flitting like ghosts across the light of their luminous neighbours, the gaunt frames of dead worlds. Here and there are vast stretches of loose cosmic dust that seems to be gathering into embryonic stars; here and there are stars in infancy or in strenuous youth. You detect all the chief phases of the making of a world in the forms and fires of these colossal aggregations of matter. Like the chance crowd on which you may look down in the square of a great city, they range from the infant to the worn and sinking aged. There is this difference, however, that the embryos of worlds sprawl, gigantic and luminous, across the expanse; that the dark and mighty bodies of the dead rush, like the rest, at twenty or fifty miles a second; and that at intervals some appalling blaze, that dims even the fearful furnaces of the living, seems to announce the resurrection of the dead. And there is this further difference, that, strewn about the intermediate space between the gigantic spheres, is a mass of cosmic dust—minute grains, or large blocks, or shoals consisting of myriads of pieces, or immeasurable clouds of fine gas—that seems to be the rubbish left over after the making of worlds, or the material gathering for the making of other worlds.

This is the universe that the nineteenth century discovered and the twentieth century is interpreting. Before we come to tell the fortunes of our little earth we have to see how matter is gathered into these stupendous globes of fire, how they come sometimes to have smaller bodies circling round them on which living things may appear, how they supply the heat and light and electricity that the living things need, and how the story of life on a planet is but a fragment of a larger story. We have to study the birth and death of worlds, perhaps the most impressive of all the studies that modern science offers us. Indeed, if we would read the whole story of evolution, there is an earlier chapter even than this; the latest chapter to be opened by science, the first to be read. We have to ask where the matter, which we are going to gather into worlds, itself came from; to understand more clearly what is the relation to it of the forces or energies—gravitation, electricity, etc.—with which we glibly mould it into worlds, or fashion it into living things; and, above all, to find out its relation to this mysterious ocean of ether in which it is found.

Less than half a century ago the making of worlds was, in popular expositions of science, a comparatively easy business. Take an indefinite number of atoms of various gases and metals, scatter them in a fine cloud over some thousands of millions of miles of space, let gravitation slowly compress the cloud into a globe, its temperature rising through the compression, let it throw off a ring of matter, which in turn gravitation

14

will compress into a globe, and you have your earth circulating round the sun. It is not quite so simple; in any case, serious men of science wanted to know how these convenient and assorted atoms happened to be there at all, and what was the real meaning of this equally convenient gravitation. There was a greater truth than he knew in the saying of an early physicist, that the atom had the look of a "manufactured article." It was increasingly felt, as the nineteenth century wore on, that the atoms had themselves been evolved out of some simpler material, and that ether might turn out to be the primordial chaos. There were even those who felt that ether would prove to be the one source of all matter and energy. And just before the century closed a light began to shine in those deeper abysses of the submaterial world, and the foundations of the universe began to appear.

CHAPTER II. THE FOUNDATIONS OF THE UNIVERSE

To the mind of the vast majority of earlier observers the phrase "foundations of the universe" would have suggested something enormously massive and solid. From what we have already seen we are prepared, on the contrary, to pass from the inconceivably large to the inconceivably small. Our sun is, as far as our present knowledge goes, one of modest dimensions. Arcturus and Canopus must be thousands of times larger than it. Yet our sun is 320,000 times heavier than the earth, and the earth weighs some 6,000,000,000,000,000,000,000,000 tons. But it is only in resolving these stupendous masses into their tiniest elements that we can reach the ultimate realities, or foundations, of the whole.

Modern science rediscovered the atoms of Democritus, analysed the universe into innumerable swarms of these tiny particles, and then showed how the infinite variety of things could be built up by their combinations. For this it was necessary to suppose that the atoms were not all alike, but belonged to a large number of different classes. From twenty-six letters of the alphabet we could make millions of different words. From forty or fifty different "elements" the chemist could construct the most varied objects in nature, from the frame of a man to a landscape. But improved methods of research led to the discovery of new elements, and at last the chemist found that he had seventy or eighty of these "ultimate realities," each having its own very definite and very different characters. As it is the experience of science to find unity underlying variety, this was profoundly unsatisfactory, and the search began for the great unity which underlay the atoms of matter. The difficulty of the search may be illustrated by a few figures. Very delicate methods were invented for calculating the size of the atoms. Laymen are apt to smile—it is a very foolish smile—at these figures, but it is enough to say that the independent and even more delicate methods suggested by recent progress in physics have quite confirmed them.

Take a cubic millimetre of hydrogen. As a millimetre is less than 1/25th of an inch, the reader must imagine a tiny bubble of gas that would fit comfortably inside the letter "o" as it is printed here. The various refined methods of the modern physicist show that there are 40,000 billion molecules (each consisting of two atoms of the gas) in this tiny bubble. It is a little universe, repeating on an infinitesimal scale the numbers and energies of the stellar universe. These molecules are not packed together, moreover, but are separated from each other by spaces which are enormous in proportion to the size of the atoms. Through these empty spaces the atoms dash at an average speed of more than a thousand miles an hour, each passing something like 6,000,000,000 of its neighbours in the course of every second. Yet this particle of gas is a thinly populated world in comparison with a particle of metal. Take a cubic centimetre of copper. In that very small square of solid matter (each side of the cube measuring a little more than a third of an inch) there are about a quadrillion atoms. It is these minute and elusive particles that modern physics sets out to master.

At first it was noticed that the atom of hydrogen was the smallest or lightest of all, and the other atoms seemed to be multiples of it. A Russian chemist, Mendeleeff, drew up a table of the elements in illustration of this, grouping them in families, which seemed to point to hydrogen as the common parent, or ultimate constituent, of each. When newly discovered elements fell fairly into place in this scheme the idea was somewhat confidently advanced that the evolution of the elements was discovered. Thus an atom of carbon seemed to be a group of 12 atoms of hydrogen, an atom of oxygen 16, an atom of sulphur 32, an atom of copper 64, an atom of silver 108, an atom of gold 197, and so on. But more correct measurements showed that these figures were not quite exact, and the fraction of inexactness killed the theory.

Long before the end of the nineteenth century students were looking wistfully to the ether for some explanation of the mystery. It was the veiled statue of Isis in the scientific world, and it resolutely kept its veil in spite of all progress. The "upper and limpid air" of the Greeks, the cosmic ocean of Giordano Bruno, was now an established reality. It was the vehicle that bore the terrific streams of energy from star to planet across the immense reaches of space. As the atoms of matter lay in it, one thought of the crystal forming in its mother-lye, or the star forming in the nebula, and wondered whether the atom was not in some such way condensed out of the ether. By the last decade of the century the theory was confidently advanced—notably by Lorentz and Larmor—though it was still without a positive basis. How the basis was found, in the last decade of the nineteenth century, may be told very briefly.

Sir William Crookes had in 1874 applied himself to the task of creating something more nearly like a vacuum than the old air-pumps afforded. When he had found the means of reducing the quantity of gas in a tube until it was a million times thinner than the atmosphere, he made the experiment of sending an electric discharge through it, and found a very curious result. From the cathode (the negative electric point) certain rays proceeded which caused a green fluorescence on the glass of the tube. Since the discharge did not consist of the atoms of the gas, he concluded that it was a new and mysterious substance, which he called "radiant matter." But no progress was made in the interpretation of this strange material. The Crookes tube became one of the toys of science—and the lamp of other investigators.

In 1895 Rontgen drew closer attention to the Crookes tube by discovering the rays which he called X-rays, but which now bear his name. They differ from ordinary light-waves in their length, their irregularity, and especially their power to pass through opaque bodies. A number of distinguished physicists now took up the study of the effect of sending an electric discharge through a vacuum, and the particles of "radiant matter" were soon identified. Sir J. J. Thomson, especially, was brilliantly successful in his interpretation. He proved that they were tiny corpuscles, more than a thousand times smaller than the atom of hydrogen, charged with negative electricity, and travelling at the rate of thousands of miles a second. They were the "electrons" in which modern physics sees the long-sought constituents of the atom.

No sooner had interest been thoroughly aroused than it was announced that a fresh discovery had opened a new shaft into the underworld. Sir J. J. Thomson, pursuing his research, found in 1896 that compounds of uranium sent out rays that could penetrate black paper and affect the photographic plate; though in this case the French physicist, Becquerel, made the discovery simultaneously' and was the first to publish it. An army of investigators turned into the new field, and sought to penetrate the deep abyss that had almost suddenly disclosed itself. The quickening of astronomy by Galilei, or of zoology by Darwin, was slight in comparison with the stirring of our physical world by these increasing discoveries. And in 1898 M. and Mme. Curie made the further discovery which, in the popular mind, obliterated all the earlier achievements. They succeeded in isolating the new element, radium, which exhibits the actual process of an atom parting with its minute constituents.

The story of radium is so recent that a few lines will suffice to recall as much as is needed for the purpose of this chapter. In their study of the emanations from uranium compounds the Curies were led to isolate the various elements of the compounds until they discovered that the discharge was predominantly due to one specific element, radium. Radium is itself probably a product of the disintegration of uranium, the heaviest of known metals, with an atomic weight some 240 times greater than that of hydrogen. But this massive atom of uranium has a life that is computed in thousands of millions of years. It is in radium and its offspring that we see most clearly the constitution of matter.

A gramme (less than 15 1/2 grains) of radium contains—we will economise our space—4×10^{21} atoms. This tiny mass is, by its discharge, parting with its substance at the rate of one atom per second for every 10,000,000,000 atoms; in other words, the "indestructible" atom has, in this case, a term of life not exceeding 2500 years. In the discharge from the radium three elements have been distinguished. The first consists of atoms of the gas helium, which are hurled off at between 10,000 and 20,000 miles a second. The third element (in the order of classification) consists of waves analogous to the Rontgen rays. But the second element is a stream of electrons, which are expelled from the atom at the appalling speed of about 100,000 miles a second. Professor Le Bon has calculated that it would take 340,000 barrels of powder to discharge a bullet at that speed. But we shall see more presently of the enormous energy displayed within the little system of the atom. We may add that after its first transformation the radium passes, much more quickly, through a further series of changes. The frontiers of the atomic systems were breaking down.

The next step was for students (notably Soddy and Rutherford) to find that radio-activity, or spontaneous discharge out of the atomic systems, was not confined to radium. Not only are other rare metals conspicuously active, but it is found that such familiar surfaces as damp cellars, rain, snow, etc., emit a lesser discharge. The value of the new material thus provided for the student of physics may be shown by one illustration. Sir J. J. Thomson observes that before these recent discoveries the investigator could not detect a gas unless about a billion molecules of it were present, and it must be remembered that the spectroscope had already gone far beyond ordinary chemical analysis in detecting the presence of substances in minute quantities. Since these discoveries we can recognise a single molecule, bearing an electric charge.

With these extraordinary powers the physicist is able to penetrate a world that lies immeasurably below the range of the most powerful microscope, and introduce us to systems more bewildering than those of the astronomer. We pass from a portentous Brobdingnagia to a still more portentous Lilliputia. It has been ascertained that the mass of the electron is the 1/1700th part of that of an atom of hydrogen, of which, as we saw, billions of molecules have ample space to execute their terrific movements within the limits of the letter "o." It has been further shown that these electrons are identical, from whatever source they are obtained. The physicist therefore concludes—warning us that on this further point he is drawing a theoretical conclusion— that the atoms of ordinary matter are made up of electrons. If that is the case, the hydrogen atom, the lightest of all, must be a complex system of some 1700 electrons, and as we ascend the scale of atomic weight the clusters grow larger and larger, until we come to the atoms of the heavier metals with more than 250,000 electrons in each atom.

But this is not the most surprising part of the discovery. Tiny as the dimensions of the atom are, they afford a vast space for the movement of these energetic little bodies. The speed of the stars in their courses is slow compared with the flight of the electrons. Since they fly out of the system, in the conditions we have described, at a speed of between 90,000 and 100,000 miles a second, they must be revolving with terrific rapidity within it. Indeed, the most extraordinary discovery of all is that of the energy imprisoned within these tiny systems, which men have for ages regarded as "dead" matter. Sir J. J. Thomson calculates that, allowing only one electron to each atom in a gramme of hydrogen, the tiny globule of gas will contain as much energy as would be obtained by burning thirty-five tons of coal. If, he says, an appreciable fraction of the energy that is contained in ordinary matter were to be set free, the earth would explode and return to its primitive nebulous condition. Mr. Fournier d'Albe tells us that the force with which electrons repel each other is a quadrillion times greater than the force of gravitation that brings atoms together; and that if two grammes of pure electrons could be placed one centimetre apart they would repel each other with a force equal to 320 quadrillion tons. The inexpert imagination reels, but it must be remembered that the speed of the electron is a measured quantity, and it is within the resources of science to estimate the force necessary to project it at that speed. [*]

* See Sir J. J. Thomson, "The Corpuscular Theory of Matter"
    (1907) and—for a more elementary presentment—"Light
    Visible and Invisible" (1911); and Mr. Fournier d'Albe, "The
    Electron Theory" (2nd. ed., 1907).

Such are the discoveries of the last fifteen years and a few of the mathematical deductions from them. We are not yet in a position to say positively that the atoms are composed of electrons, but it is clear that the experts are properly modest in claiming only that this is highly probable. The atom seems to be a little universe in which, in combination with positive electricity (the nature of which is still extremely obscure), from 1700 to 300,000 electrons revolve at a speed that reaches as high as 100,000 miles a second. Instead of being crowded together, however, in their minute system, each of them has, in proportion to its size, as ample a space to move in as a single speck of dust would have in a moderate-sized room (Thomson). This theory not only meets all the facts that have been discovered in an industrious decade of research, not only offers a splendid prospect of introducing unity into the eighty-one different elements of the chemist, but it opens out a still larger prospect of bringing a common measure into the diverse forces of the universe.

Light is already generally recognised as a rapid series of electro-magnetic waves or pulses in ether. Magnetism becomes intelligible as a condition of a body in which the electrons revolve round the atom in nearly the same plane. The difference between positive and negative electricity is at least partly illuminated. An atom will repel an atom when its equilibrium is disturbed by the approach of an additional electron; the physicist even follows the movement of the added electron, and describes it revolving 2200 billion times a second round the atom, to escape being absorbed in it. The difference between good and bad conductors of electricity becomes intelligible. The atoms of metals are so close together that the roaming electrons pass freely from one atom to another, in copper, it is calculated, the electron combines with an atom and is liberated again a hundred million times a second. Even chemical action enters the sphere of explanation.

However these hypotheses may fare, the electron is a fact, and the atom is very probably a more or less stable cluster of electrons. But when we go further, and attempt to trace the evolution of the electron out of ether, we enter a region of pure theory. Some of the experts conceive the electron as a minute whirlpool or vortex in the ocean of ether; some hold that it is a centre of strain in ether; some regard ether as a densely packed mass of infinitely small grains, and think that the positive and negative corpuscles, as they seem to us, are tiny areas in which the granules are unequally distributed. Each theory has its difficulties. We do not know the origin of the electron, because we do not know the nature of ether. To some it is an elastic solid, quivering in waves at every movement of the particles; to others it is a continuous fluid, every cubic millimetre of which possesses "an energy equivalent to the output of a million-horse-power station for 40.000,000 years" (Lodge); to others it is a close-packed granular mass with a pressure of 10,000 tons per square centimetre. We must wait. It is little over ten years since the vaults were opened and physicists began to peer into the sub-material world. The lower, perhaps lowest, depth is reserved for another generation.

But it may be said that the research of the last ten years has given us a glimpse of the foundations of the universe. Every theory of the electron assumes it to be some sort of nodule or disturbed area in the ether. It is sometimes described as "a particle of negative electricity" and associated with "a particle of positive electricity" in building up the atom. The phrase is misleading for those who regard electricity as a force or energy, and it gives rise to speculation as to whether "matter" has not been resolved into "force." Force or energy is not conceived by physicists as a substantial reality, like matter, but an abstract expression of certain relations of matter or electrons.

In any case, the ether, whether solid or fluid or granular, remains the fundamental reality. The universe does not float IN an ocean of ether: it IS an ocean of ether. But countless myriads of minute disturbances are found in this ocean, and set it quivering with the various pulses which we classify as forces or energies. These points of disturbance cluster together in systems (atoms) of from 1000 to 250,000 members, and the atoms are pressed together until they come in the end to form massive worlds. It remains only to reduce gravitation itself, which brings the atoms together, to a strain or stress in ether, and we have a superb unity. That has not yet been done, but every theory of gravitation assumes that it is a stress in the ether corresponding to the formation of the minute disturbances which we call electrons.

But, it may be urged, he who speaks of foundations speaks of a beginning of a structure; he who speaks of evolution must have a starting-point. Was there a time when the ether was a smooth, continuous fluid, without electrons or atoms, and did they gradually appear in it, like crystals in the mother-lye? In science we know nothing of a beginning. The question of the eternity or non-eternity of matter (or ether) is as futile as the question about its infinity or finiteness. We shall see in the next chapter that science can trace the processes of nature back for hundreds, if not thousands, of millions of years, and has ground to think that the universe then presented much the same aspect as it does now, and will in thousands of millions of years to come. But if these periods were quadrillions, instead of millions, of years, they would still have no relation to the idea of eternity. All that we can say is that we find nothing in nature that points to a beginning or an end. [*]

* A theory has been advanced by some physicists that there
is evidence of a beginning. WITHIN OUR EXPERIENCE energy is
being converted into heat more abundantly than heat is being
converted into other energy. This would hold out a prospect
of a paralysed universe, and that stage would have been
reached long ago if the system had not had a definite
beginning. But what knowledge have we of conversions of
energy in remote regions of space, in the depths of stars or
nebulae, or in the sub-material world of which we have just
caught a glimpse? Roundly, none. The speculation is
worthless.

One point only need be mentioned in conclusion. Do we anywhere perceive the evolution of the material elements out of electrons, just as we perceive the devolution, or disintegration, of atoms into electrons? There is good ground for thinking that we do. The subject will be discussed more fully in the next chapter. In brief, the spectroscope, which examines the light of distant stars and discovers what chemical elements emitted it, finds matter, in the hottest stars, in an unusual condition, and seems to show the elements successively emerging from their fierce alchemy. Sir J. Norman Lockyer has for many years conducted a special investigation of the subject at the Solar Physics Observatory, and he declares that we can trace the evolution of the elements out of the fiery chaos of the young star. The lightest gases emerge first, the metals later, and in a special form. But here we pass once more from Lilliputia to Brobdingnagia, and must first explain the making of the star itself.

CHAPTER III. THE BIRTH AND DEATH OF WORLDS

The greater part of this volume will be occupied with the things that have happened on one small globe in the universe during a certain number of millions of years. It cannot be denied that this has a somewhat narrow and parochial aspect. The earth is, you remember, a million times smaller than the sun, and the sun itself is a very modest citizen of the stellar universe. Our procedure is justified, however, both on the ground of personal interest, and because our knowledge of the earth's story is so much more ample and confident. Yet we must preface the story of the earth with at least a general outline of the larger story of the universe. No sensible man is humbled or dismayed by the vastness of the universe. When the human mind reflects on its wonderful scientific mastery of this illimitable ocean of being, it has no sentiment of being dwarfed or degraded. It looks out with cold curiosity over the mighty scattering of worlds, and asks how they, including our own world, came into being.

We now approach this subject with a clearer perception of the work we have to do. The universe is a vast expanse of ether, and somehow or other this ether gives rise to atoms of matter. We may imagine it as a spacious chamber filled with cosmic dust; recollecting that the chamber has no walls, and that the dust arises in the ether itself. The problem we now approach is, in a word: How are these enormous stretches of cosmic dust, which we call matter, swept together and compressed into suns and planets? The most famous answer to this question is the "nebular hypothesis." Let us see, briefly, how it came into modern science.

We saw that some of the ancient Greek speculators imagined their infinite number of atoms as scattered originally, like dust, throughout space and gradually coming together, as dust does, to form worlds. The way in which they brought their atoms together was wrong, but the genius of Democritus had provided the germ of another sound theory to the students of a more enlightened age. Descartes (1596-1650) recalled the idea, and set out a theory of the evolution of stars and planets from a diffused chaos of particles. He even ventured to say that the earth was at one time a small white-hot sun, and that a solid crust had gradually formed round its molten core. Descartes had taken refuge in Sweden from his persecutors, and it is therefore not surprising that that strange genius Swedenborg shortly afterwards developed the same idea. In the middle of the eighteenth century the great French naturalist, Buffon, followed and improved upon Descartes and Swedenborg. From Buffon's work it was learned by the German philosopher Kant, who published (1755) a fresh theory of the concentration of scattered particles into fiery worlds. Then Laplace (1749-1827) took up the speculation, and gave it the form in which it practically ruled astronomy throughout the nineteenth century. That is the genealogy of the famous nebular hypothesis. It did not spring full-formed from the brain of either Kant or Laplace, like Athene from the brain of Zeus.

Laplace had one great advantage over the early speculators. Not only was he an able astronomer and mathematician, but by his time it was known that nebulae, or vast clouds of dispersed matter, actually existed in the heavens. Here was a solid basis for the speculation. Sir William Herschel, the most assiduous explorer of the heavens, was a contemporary of Laplace. Laplace therefore took the nebula as his starting-point.

A quarter of an ounce of solid matter (say, tobacco) will fill a vast space when it is turned into smoke, and if it were not for the pressure of the atmosphere it would expand still more. Laplace imagined the billions of tons of matter which constitute our solar system similarly dispersed, converted into a fine gas, immeasurably thinner than the atmosphere. This nebula would be gradually drawn in again by gravitation, just as the dust falls to the floor of a room. The collisions of its particles as they fell toward the centre would raise its temperature and give it a rotating movement. A time would come when the centrifugal force at the outer ring of the rotating disk would equal the centripetal (or inward) pull of gravity, and this ring would be detached, still spinning round the central body. The material of the ring would slowly gather, by gravitation, round some denser area in it; the ring would become a sphere; we should have the first, and outermost, planet circling round the sun. Other rings would successively be detached, and form the rest of the planets; and the sun is the shrunken and condensed body of the nebula.

So simple and beautiful a theory of the solar system could not fail to captivate astronomers, but it is generally rejected to-day, in the precise form which Laplace gave it. What the difficulties are which it has encountered, and the modifications it must suffer, we shall see later; as well as the new theories which have largely displaced it. It will be better first to survey the universe from the evolutionary point of view. But I may observe, in passing, that the sceptical remarks one hears at times about scientific theories contradicting and superseding each other are frivolous. One great idea pervades all the theories of the evolution of worlds, and that idea is firmly established. The stars and their planets are enormous aggregations of cosmic dust, swept together and compressed by the action of gravitation. The precise nature of this cosmic dust—whether it was gas, meteorites and gas, or other particles—is open to question.

As we saw in the first chapter, the universe has the word evolution written, literally, in letters of fire across it. The stars are of all ages, from sturdy youth to decrepit age, and even to the darkness of death. We saw that this can be detected on the superficial test of colour. The colours of the stars are, it is true, an unsafe ground to build upon. The astronomer still puzzles over the gorgeous colours he finds at times, especially in double stars: the topaz and azure companions in beta Cygni, the emerald and red of alpha Herculis, the yellow and rose of eta Cassiopeiae, and so on. It is at the present time under discussion in astronomy how far these colours are objective at all, or whether, if they are real, they may not be due to causes other than temperature. Yet the significance of the three predominating colours—blue-white, yellow, and red— has been sustained by the spectroscope. It is the series of colours through which a white-hot bar of iron passes as it cools. And the spectroscope gives us good ground to conclude that the stars are cooling.

When a glowing gas (not under great pressure) is examined by the spectroscope, it yields a few vertical lines or bars of light on a dark background; when a glowing liquid or solid is examined, it gives a continuous rainbow-like stretch of colour. Some of the nebulae give the former type of spectrum, and are thus known to be masses of luminous gas; many of the nebulae and the stars have the latter type of spectrum. But the stretch of light in the spectrum of a star is crossed, vertically, by a number of dark lines, and experiment in the laboratory has taught us how to interpret these. They mean that there is some light-absorbing vapour between the source of light and the instrument. In the case of the stars they indicate the presence of an atmosphere of relatively cool vapours, and an increase in the density of that atmosphere—which is shown by a multiplication and broadening of the dark lines on the spectrum—means an increase of age, a loss of vitality, and ultimately death. So we get the descending scale of spectra. The dark lines are thinnest and least numerous in the blue stars, more numerous in the yellow, heavy and thick in the red. As the body of the star sinks in temperature dense masses of cool vapour gather about it. Its light, as we perceive it, turns yellow, then red. The next step, which the spectroscope cannot follow, will be the formation of a scum on the cooling surface, ending, after ages of struggle, in the imprisonment of the molten interior under a solid, dark crust. Let us see how our sun illustrates this theory.

It is in the yellow, or what we may call the autumnal, stage. Miss Clerke and a few others have questioned this, but the evidence is too strong to-day. The vast globe, 867,000 miles in diameter, seems to be a mass of much the same material as the earth—about forty elements have been identified in it—but at a terrific temperature. The light-giving surface is found, on the most recent calculations, to have a temperature of about 6700 degrees C. This surface is an ocean of liquid or vaporised metals, several thousand miles in depth; some think that the brilliant light comes chiefly from clouds of incandescent carbon. Overlying it is a deep layer of the vapours of the molten metals, with a temperature of about 5500 degrees C.; and to this comparatively cool and light-absorbing layer we owe the black lines of the solar spectrum. Above it is an ocean of red-hot hydrogen, and outside this again is an atmosphere stretching for some hundreds of thousands of miles into space.

The significant feature, from our point of view, is the "sun-spot"; though the spot may be an area of millions of square miles. These areas are, of course, dark only by comparison with the intense light of the rest of the disk. The darkest part of them is 5000 times brighter than the full moon. It will be seen further, on examining a photograph of the sun, that a network or veining of this dark material overspreads the entire surface at all times. There is still some difference of opinion as to the nature of these areas, but the evidence of the spectroscope has convinced most astronomers that they are masses of cooler vapour lying upon, and sinking into, the ocean of liquid fire. Round their edges, as if responding to the pressure of the more condensed mass, gigantic spurts and mountains of the white-hot matter of the sun rush upwards at a rate of fifty or a hundred miles a second, Sometimes they reach a height of a hundred, and even two hundred, thousand miles, driving the red-hot hydrogen before them in prodigious and fantastic flames. Between the black veins over the disk, also, there rise domes and columns of the liquid fire, some hundreds of miles in diameter, spreading and sinking at from five to twenty miles a second. The surface of the sun—how much more the interior!—is an appalling cauldron of incandescent matter from pole to pole. Every yard of the surface is a hundred times as intense as the open furnace of a Titanic. From the depths and from the surface of this fiery ocean, as, on a small scale, from the surface of the tropical sea, the vapours rise high into the extensive atmosphere, discharge some of their heat into space, and sink back, cooler and heavier, upon the disk.

This is a star in its yellow age, as are Capella and Arcturus and other stars. The red stars carry the story further, as we should expect. The heavier lines in their spectrum indicate more absorption of light, and tell us that the vapours are thickening about the globe; while compounds like titanium oxide make their appearance, announcing a fall of temperature. Below these, again, is a group of dark red or "carbon" stars, in which the process is carried further. Thick, broad, dark lines in the red end of the spectrum announce the appearance of compounds of carbon, and a still lower fall of temperature. The veil is growing thicker; the life is ebbing from the great frame. Then the star sinks below the range of visibility, and one would think that we can follow the dying world no farther. Fortunately, in the case of Algol and some thirty or forty other stars, an extinct sun betrays its existence by flitting across the light of a luminous sun, and recent research has made it probable that the universe is strewn with dead worlds. Some of them may be still in the condition which we seem to find in Jupiter, hiding sullen fires under a dense shell of cloud; some may already be covered with a crust, like the earth. There are even stars in which one is tempted to see an intermediate stage: stars which blaze out periodically from dimness, as if the Cyclops were spending his last energy in spasms that burst the forming roof of his prison. But these variable stars are still obscure, and we do not need their aid. The downward course of a star is fairly plain.

When we turn to the earlier chapters in the life of a star, the story is less clear. It is at least generally agreed that the blue-white stars exhibit an earlier and hotter stage. They show comparatively little absorption, and there is an immense preponderance of the lighter gases, hydrogen and helium. They (Sirius, Vega, etc.) are, in fact, known as "hydrogen stars," and their temperature is generally computed at between 20,000 and 30,000 degrees C. A few stars, such as Procyon and Canopus, seem to indicate a stage between them and the yellow or solar type. But we may avoid finer shades of opinion and disputed classes, and be content with these clear stages. We begin with stars in which only hydrogen and helium, the lightest Of elements, can be traced; and the hydrogen is in an unfamiliar form, implying terrific temperature. In the next stage we find the lines of oxygen, nitrogen, magnesium, and silicon. Metals such as iron and copper come later, at first in a primitive and unusual form. Lastly we get the compounds of titanium and carbon, and the densely shaded spectra which tell of the thickly gathering vapours. The intense cold of space is slowly prevailing in the great struggle.

What came before the star? It is now beyond reasonable doubt that the nebula—taking the word, for the moment, in the general sense of a loose, chaotic mass of material—was the first stage. Professor Keeler calculated that there are at least 120,000 nebulae within range of our telescopes, and the number is likely to be increased. A German astronomer recently

36

counted 1528 on one photographic plate. Many of them, moreover, are so vast that they must contain the material for making a great number of worlds. Examine a good photograph of the nebula in Orion. Recollect that each one of the points of light that are dotted over the expanse is a star of a million miles or more in diameter (taking our sun as below the average), and that the great cloud that sprawls across space is at least 10,000 billion miles away; how much more no man knows. It is futile to attempt to calculate the extent of that vast stretch of luminous gas. We can safely say that it is at least a million times as large as the whole area of our solar system; but it may run to trillions or quadrillions of miles.

Nearly a hundred other nebulae are known, by the spectroscope, to be clouds of luminous gas. It does not follow that they are white-hot, and that the nebula is correctly called a "fire-mist." Electrical and other agencies may make gases luminous, and many astronomers think that the nebulae are intensely cold. However, the majority of the nebulae that have been examined are not gaseous, and have a very different structure from the loose and diffused clouds of gas. They show two (possibly more, but generally two) great spiral arms starting from the central part and winding out into space. As they are flat or disk-shaped, we see this structure plainly when they turn full face toward the earth, as does the magnificent nebula in Canes Venatici. In it, and many others, we clearly trace a condensed central mass, with two great arms, each apparently having smaller centres of condensation, sprawling outward like the broken spring of a watch.

The same structure can be traced in the mighty nebula in Andromeda, which is visible to the naked eye, and it is said that more than half the nebulae in the heavens are spiral. Knowing that they are masses of solid or liquid fire, we are tempted to see in them gigantic Catherine-wheels, the fireworks of the gods. What is their relation to the stars?

In the first place, their mere existence has provided a solid basis for the nebular hypothesis, and their spiral form irresistibly suggests that they are whirling round on their central axis and concentrating. Further, we find in some of the gaseous nebulae (Orion) comparatively void spaces occupied by stars, which seem to have absorbed the nebulous matter in their formation. On the other hand, we find (in the Pleiades) wisps and streamers of nebulous matter clinging about great clusters of stars, suggesting that they are material left over when these clustered worlds crystallised out of some vast nebula; and enormous stretches of nebulous material covering regions (as in Perseus) where the stars are as thick as grains of silver. More important still, we find a type of cosmic body which seems intermediate between the star and the nebula. It is a more or less imperfectly condensed star, surrounded by nebular masses. But one of the most instructive links of all is that at times a nebula is formed from a star, and a recent case of this character may be briefly described.

In February, 1901, a new star appeared in the constellation Perseus. Knowing what a star is, the reader will have some dim conception of the portentous blaze that lit up that remote region of space (at least 600 billion miles away) when he learns that the light of this star increased 4000-fold in twenty-eight hours. It reached a brilliance 8000 times greater than that of the sun. Telescopes and spectroscopes were turned on it from all parts of the earth, and the spectroscope showed that masses of glowing hydrogen were rushing out from it at a rate of nearly a thousand miles a second. Its light gradually flickered and fell, however, and the star sank back into insignificance. But the photographic plate now revealed a new and most instructive feature. Before the end of the year there was a nebula, of enormous extent, spreading out on both sides from the centre of the eruption. It was suggested at the time that the bursting of a star may merely have lit up a previously dark nebula, but the spectroscope does not support this. A dim star had dissolved, wholly or partially, into a nebula, as a result of some mighty cataclysm. What the nature of the catastrophe was we will inquire presently.

These are a few of the actual connections that we find between stars and nebulae. Probably, however, the consideration that weighs most with the astronomer is that the condensation of such a loose, far-stretched expanse of matter affords an admirable explanation of the enormous heat of the stars. Until recently there was no other conceivable source that would supply the sun's tremendous outpour of energy for tens of millions of years except the compression of its substance. It is true that the discovery of radio-activity has disclosed a new source of energy within the atoms themselves, and there are scientific men, like Professor Arrhenius, who attach great importance to this source. But, although it may prolong the limited term of life which physicists formerly allotted to the sun and other stars, it is still felt that the condensation of a nebula offers the best explanation of the origin of a sun, and we have ample evidence for the connection. We must, therefore, see what the nebula is, and how it develops.

"Nebula" is merely the Latin word for cloud. Whatever the nature of these diffused stretches of matter may be, then, the name applies fitly to them, and any theory of the development of a star from them is still a "nebular hypothesis." But the three theories which divide astronomers to-day differ as to the nature of the nebula. The older theory, pointing to the gaseous nebulae as the first stage, holds that the nebula is a cloud of extremely attenuated gas. The meteoritic hypothesis (Sir N. Lockyer, Sir G. Darwin, etc.), observing that space seems to swarm with meteors and that the greater part of the nebulae are not gaseous, believes that the starting-point is a colossal swarm of meteors, surrounded by the gases evolved and lit up by their collisions. The planetesimal hypothesis, advanced in recent years by Professor Moulton and Professor Chamberlin, contends that the nebula is a vast cloud of liquid or solid (but not gaseous) particles. This theory is based mainly on the dynamical difficulties of the other two, which we will notice presently.

The truth often lies between conflicting theories, or they may apply to different cases. It is not improbable that this will be our experience in regard to the nature of the initial nebula. The gaseous nebulae, and the formation of such nebulae from disrupted stars, are facts that cannot be ignored. The nebulae with a continuous spectrum, and therefore—in part, at least—in a liquid or solid condition, may very well be regarded as a more advanced stage of condensation of the same; their spiral shape and conspicuous nuclei are consistent with this. Moreover, a condensing swarm of meteors would, owing to the heat evolved, tend to pass into a gaseous condition. On the tether hand, a huge expanse of gas stretched over billions of miles of space would be a net for the wandering particles, meteors, and comets that roam through space. If it be true, as is calculated, that our 24,000 miles of atmosphere capture a hundred million meteors a day, what would the millions or billions of times larger net of a nebula catch, even if the gas is so much thinner? In other words, it is not wise to draw too fine a line between a gaseous nebula and one consisting of solid particles with gas.

The more important question is: How do astronomers conceive the condensation of this mixed mass of cosmic dust? It is easy to reply that gravitation, or the pressure of the surrounding ether, slowly drives the particles centre-ward, and compresses the dust into globes, as the boy squeezes the flocculent snow into balls; and it is not difficult for the mathematician to show that this condensation would account for the shape and temperature of the stars. But we must go a little beyond this superficial statement, and see, to some extent, how the deeper students work out the process. [*]

   * See, especially, Dr. P. Lowell, "The Evolution of Worlds"
      (1909). Professor S. Arrhenius, "Worlds in the Making"

(1908), Sir N. Lockyer, "The Meteorite Hypothesis" (1890),

Sir R. Ball, "The Earth's Beginning" (1909), Professor

Moulton, "The Astrophysical Journal (October, 1905), and

Chamberlin and Salisbury, "Geology," Vol. II. (1903).

Taking a broad view of the whole field, one may say that the two chief difficulties are as follows: First, how to get the whole chaotic mass whirling round in one common direction; secondly, how to account for the fact that in our solar system the outermost planets and satellites do not rotate in the same direction as the rest. There is a widespread idea that these difficulties have proved fatal to the old nebular hypothesis, and there are distinguished astronomers who think so. But Sir R. Ball (see note), Professor Lowell (see note), Professor Pickering (Annals of Harvard College Observatory, 53, III), and other high authorities deny this, and work out the newly discovered movements on the lines of the old theory. They hold that all the bodies in the solar system once turned in the same direction as Uranus and Neptune, and the tidal influence of the sun has changed the rotation of most of them. The planets farthest from the sun would naturally not be so much affected by it. The same principle would explain the retrograde movement of the outer satellites of Saturn and Jupiter. Sir R. Ball further works out the principles on which the particles of the condensing nebula would tend to form a disk rotating on its central axis. The ring-theory of Laplace is practically abandoned. The spiral nebula is evidently the standard type, and the condensing nebula must conform to it. In this we are greatly helped by the current theory of the origin of spiral nebulae.

We saw previously that new stars sometimes appear in the sky, and the recent closer scrutiny of the heavens shows this occurrence to be fairly frequent. It is still held by a few astronomers that such a cataclysm means that two stars collided. Even a partial or "grazing" collision between two masses, each weighing billions of tons, travelling (on the average) forty or fifty miles a second—a movement that would increase enormously as they approach each other—would certainly liquefy or vaporise their substance; but the astronomer, accustomed to see cosmic bodies escape each other by increasing their speed, is generally disinclined to believe in collisions. Some have made the new star plunge into the heart of a dense and dark nebula; some have imagined a shock of two gigantic swarms of meteors; some have regarded the outflame as the effect of a prodigious explosion. In one or other new star each or any of these things may have occurred, but the most plausible and accepted theory for the new star of 1901 and some others is that two stars had approached each other too closely in their wandering. Suppose that, in millions of years to come, when our sun is extinct and a firm crust surrounds the great molten ball, some other sun approaches within a few million miles of it. The two would rush past each other at a terrific speed, but the gravitational effect of the approaching star would tear open the solid shell of the sun, and, in a mighty flame, its molten and gaseous entrails would be flung out into space. It has long been one of the arguments against a molten interior of the earth that the sun's gravitational influence would raise it in gigantic tides and rend the solid shell of rock. It is even suspected now that our small earth is not without a tidal influence on the sun. The comparatively near approach of two suns would lead to a terrific cataclysm.

If we accept this theory, the origin of the spiral nebula becomes intelligible. As the sun from which it is formed is already rotating on its axis, we get a rotation of the nebula from the first. The mass poured out from the body of the sun would, even if it were only a small fraction of its mass, suffice to make a planetary system; all our sun's planets and their satellites taken together amount to only 1/100th of the mass of the solar system. We may assume, further, that the outpoured matter would be a mixed cloud of gases and solid and liquid particles; and that it would stream out, possibly in successive waves, from more than one part of the disrupted sun, tending to form great spiral trails round the parent mass. Some astronomers even suggest that, as there are tidal waves raised by the moon at opposite points of the earth, similar tidal outbursts would occur at opposite points on the disk of the disrupted star, and thus give rise to the characteristic arms starting from opposite sides of the spiral nebula. This is not at all clear, as the two tidal waves of the earth are due to the fact that it has a liquid ocean rolling on, not under, a solid bed.

In any case, we have here a good suggestion of the origin of the spiral nebula and of its further development. As soon as the outbursts are over, and the scattered particles have reached the farthest limit to which they are hurled, the concentrating action of gravitation will slowly assert itself. If we conceive this gravitational influence as the pressure of the surrounding ether we get a wider understanding of the process. Much of the dispersed matter may have been shot far enough into space to escape the gravitational pull of the parent mass, and will be added to the sum of scattered cosmic dust, meteors, and close shoals of meteors (comets) wandering in space. Much of the rest will fall back upon the central body But in the great spiral arms themselves the distribution of the matter will be irregular, and the denser areas will slowly gather in the surrounding material. In the end we would thus get secondary spheres circling round a large primary.

This is the way in which astronomers now generally conceive the destruction and re-formation of worlds. On one point the new planetesimal theory differs from the other theories. It supposes that, since the particles of the whirling nebula are all travelling in the same general direction, they overtake each other with less violent impact than the other theories suppose, and therefore the condensation of the material into planets would not give rise to the terrific heat which is generally assumed. We will consider this in the next chapter, when we deal with the formation of the planets. As far as the central body, the sun, is concerned, there can be no hesitation. The 500,000,000 incandescent suns in the heavens are eloquent proof of the appalling heat that is engendered by the collisions of the concentrating particles.

In general outline we now follow the story of a star with some confidence. An internal explosion, a fatal rush into some dense nebula or swarm of meteors, a collision with another star, or an approach within a few million miles of another star, scatters, in part or whole, the solid or liquid globe in a cloud of cosmic dust. When the violent outrush is over, the dust is gathered together once more into a star. At first cold and attenuated, its temperature rises as the particles come together, and we have, after a time, an incandescent nucleus shining through a thin veil of gas—a nebulous star. The temperature rises still further, and we have the blue-hot star, in which the elements seem to be dissociated, and slowly re-forming as the temperature falls. After, perhaps, hundreds of millions of years it reaches the "yellow" stage, and, if it has planets with the conditions of life, there may be a temporary opportunity for living things to enjoy its tempered energy. But the cooler vapours are gathering round it, and at length its luminous body is wholly imprisoned. It continues its terrific course through space, until some day, perhaps, it again encounters the mighty cataclysm which will make it begin afresh the long and stormy chapters of its living history.

Such is the suggestion of the modern astronomer, and, although we seem to find every phase of the theory embodied in the varied contents of the heavens, we must not forget that it is only a suggestion. The spectroscope and telescopic photography, which are far more important than the visual telescope, are comparatively recent, and the field to be explored is enormous. The mist is lifting from the cosmic

46

landscape, but there is still enough to blur our vision. Very puzzling questions remain unanswered. What is the origin of the great gaseous nebulae? What is the origin of the triple or quadruple star? What is the meaning of stars whose light ebbs and flows in periods of from a few to several hundred days? We may even point to the fact that some, at least, of the spiral nebulae are far too vast to be the outcome of the impact or approach of two stars.

We may be content to think that we have found out some truths, by no means the whole truth, about the evolution of worlds. Throughout this immeasurable ocean of ether the particles of matter are driven together and form bodies. These bodies swarm throughout space, like fish in the sea; travelling singly (the "shooting star"), or in great close shoals (the nucleus of a comet), or lying scattered in vast clouds. But the inexorable pressure urges them still, until billions of tons of material are gathered together. Then, either from the sheer heat of the compression, or from the formation of large and unstable atomic systems (radium, etc.), or both, the great mass becomes a cauldron of fire, mantled in its own vapours, and the story of a star is run. It dies out in one part of space to begin afresh in another. We see nothing in the nature of a beginning or an end for the totality of worlds, the universe. The life of all living things on the earth, from the formation of the primitive microbes to the last struggles of the superman, is a small episode of that stupendous drama, a fraction of a single scene. But our ampler knowledge of it, and our personal interest in it, magnify that episode, and we turn from the cosmic picture to study the formation of the earth and the rise of its living population.

CHAPTER IV. THE PREPARATION OF THE EARTH

The story of the evolution of our solar system is, it will now be seen, a local instance of the great cosmic process we have studied in the last chapter. We may take one of the small spiral nebulae that abound in the heavens as an illustration of the first stage. If a still earlier stage is demanded, we may suppose that some previous sun collided with, or approached too closely, another mighty body, and belched out a large part of its contents in mighty volcanic outpours. Mathematical reasoning can show that this erupted material would gather into a spiral nebula; but, as mathematical calculations cannot be given here, and are less safe than astronomical facts, we will be content to see the early shape of our solar system in a relatively small spiral nebula, its outermost arm stretching far beyond the present orbit of Neptune, and its great nucleus being our present sun in more diffused form.

We need not now attempt to follow the shrinking of the central part of the nebula until it becomes a rounded fiery sun. That has been done in tracing the evolution of a star. Here we have to learn how the planets were formed from the spiral arms of the nebula.

The principle of their formation is already clear. The same force of gravitation, or the same pressure of the surrounding ether, which compresses the central mass into a fiery globe, will act upon the loose material of the arms and compress it into smaller globes. But there is an interesting and acute difference of opinion amongst modern experts as to whether these smaller globes, the early planets, would become white-hot bodies.

The general opinion, especially among astronomers, is that the compression of the nebulous material of the arms into globes would generate enormous heat, as in the case of the sun. On that view the various planets would begin their careers as small suns, and would pass through those stages of cooling and shrinking which we have traced in the story of the stars. A glance at the photograph of one of the spiral nebulae strongly confirms this. Great luminous knots, or nuclei, are seen at intervals in the arms. Smaller suns seem to be forming in them, each gathering into its body the neighbouring material of the arm, and rising in temperature as the mass is compressed into a globe. The spectroscope shows that these knots are condensing masses of white-hot liquid or solid matter. It therefore seems plain that each planet will first become a liquid globe of fire, coursing round the central sun, and will gradually, as its heat is dissipated and the supply begins to fail, form a solid crust.

This familiar view is challenged by the new "planetesimal hypothesis," which has been adopted by many distinguished geologists (Chamberlin, Gregory, Coleman, etc.). In their view the particles in the arms of the nebula are all moving in the same direction round the sun. They therefore quietly overtake the nucleus to which they are attracted, instead of violently colliding with each other, and much less heat is generated at the surface. In that case the planets would not pass through a white-hot, or even red-hot, stage at all. They are formed by a slow ingathering of the scattered particles, which are called "planetesimals" round the larger or denser masses of stuff which were discharged by the exploding sun. Possibly these masses were prevented from falling back into the sun by the attraction of the colliding body, or the body which caused the eruption. They would revolve round the parent body, and the shoals of smaller particles would gather about them by gravitation. If there were any large region in the arm of the nebula which had no single massive nucleus, the cosmic dust would gather about a number of smaller centres. Thus might be explained the hundreds of planetoids, or minor planets, which we find between Mars and Jupiter. If these smaller bodies came within the sphere of influence of one of the larger planets, yet were travelling quickly enough to resist its attraction, they would be compelled to revolve round it, and we could thus explain the ten satellites of Saturn and the eight of Jupiter. Our moon, we shall see, had a different origin.

We shall find this new hypothesis crossing the familiar lines at many points in the next few chapters. We will consider those further consequences as they arise, but may say at once that, while the new theory has greatly helped us in tracing the formation of the planetary system, astronomers are strongly opposed to its claim that the planets did not pass through an incandescent stage. The actual features of our spiral nebulae seem clearly to exhibit that stage. The shape of the planets—globular bodies, flattened at the poles—strongly suggests that they were once liquid. The condition in which we find Saturn and Jupiter very forcibly confirms this suggestion; the latest study of those planets supports the current opinion that they are still red-hot, and even seems to detect the glow of their surfaces in their mantles of cloud. These points will be considered more fully presently. For the moment it is enough to note that, as far as the early stages of planetary development are concerned, the generally accepted theory rests on a mass of positive evidence, while the new hypothesis is purely theoretical. We therefore follow the prevailing view with some confidence.

Those of the spiral nebulae which face the earth squarely afford an excellent suggestion of the way in which planets are probably formed. In some of these nebulae the arms consist of almost continuous streams of faintly luminous matter; in others the matter is gathering about distinct centres; in others again the nebulous matter is, for the most part, collected in large glowing spheres. They seem to be successive stages, and to reveal to us the origin of our planets. The position of each planet in our solar system would be determined by the chance position of the denser stuff shot out by the erupting sun. I have seen Vesuvius hurl up into the sky, amongst its blasts of gas and steam, white-hot masses of rock weighing fifty tons. In the far fiercer outburst of the erupting sun there would be at least thinner and denser masses, and they must have been hurled so far into space that their speed in travelling round the central body, perhaps seconded by the attraction of the second star, overcame the gravitational pull back to the centre. Recollect the force which, in the new star in Perseus, drove masses of hydrogen for millions of miles at a speed of a thousand miles a second.

These denser nuclei or masses would, when the eruption was over, begin to attract to themselves all the lighter nebulous material within their sphere of gravitational influence. Naturally, there would at first be a vast confusion of small and large centres of condensation in the arms of the nebula, moving in various directions, but a kind of natural selection— and, in this case, survival of the biggest—would ensue. The conflicting movements would be adjusted by collisions and gravitation, the smaller bodies would be absorbed in the larger or enslaved as their satellites, and the last state would be a family of smaller suns circling at vast distances round the parent body. The planets, moreover, would be caused to rotate on their axes, besides revolving round the sun, as the particles at their inner edge (nearer the sun) would move at a different speed from those at the outer edge. In the course of time the smaller bodies, having less heat to lose and less (or no) atmosphere to check the loss, would cool down, and become dark solid spheres, lit only by the central fire.

While the first stage of this theory of development is seen in the spiral nebula, the later stages seem to be well exemplified in the actual condition of our planets.

Following, chiefly, the latest research of Professor Lowell and his colleagues, which marks a considerable advance on our previous knowledge, we shall find it useful to glance at the sister-planets before we approach the particular story of our earth.

Mercury, the innermost and smallest of the planets, measuring only some 3400 miles in diameter, is, not unexpectedly, an airless wilderness. Small bodies are unable to retain the gases at their surface, on account of their feebler gravitation. We find, moreover, that Mercury always presents the same face to the sun, as it turns on its axis in the same period (eighty-eight days) in which it makes a revolution round the sun. While, therefore, one half of the globe is buried in eternal darkness, the other half is eternally exposed to the direct and blistering rays of the sun, which is only 86,000,000 miles away. To Professor Lowell it presents the appearance of a bleached and sun-cracked desert, or "the bones of a dead world." Its temperature must be at least 300 degrees C. above that of the earth. Its features are what we should expect on the nebular hypothesis. The slowness of its rotation is accounted for by the heavy tidal influence of the sun. In the same way our moon has been influenced by the earth, and our earth by the sun, in their movement of rotation.

Venus, as might be expected in the case of so large a globe (nearly as large as the earth), has an atmosphere, but it seems, like Mercury, always to present the same face to the sun. Its comparative nearness to the sun (67,000,000 miles) probably explains this advanced effect of tidal action. The consequences that the observers deduce from the fact are interesting. The sun-baked half of Venus seems to be devoid of water or vapour, and it is thought that all its water is gathered into a rigid ice-field on the dark side of the globe, from which fierce hurricanes must blow incessantly. It is a Sahara, or a desert far hotter than the Sahara, on one side; an arctic region on the other. It does not seem to be a world fitted for the support of any kind of life that we can imagine.

When we turn to the consideration of Mars, we enter a world of unending controversy. With little more than half the diameter of the earth, Mars ought to be in a far more advanced stage of either life or decay, but its condition has not yet been established. Some hold that it has a considerable atmosphere; others that it is too small a globe to have retained a layer of gas. Professor Poynting believes that its temperature is below the freezing-point of water all over the globe; many others, if not the majority of observers, hold that the white cap we see at its poles is a mass of ice and snow, or at least a thick coat of hoar-frost, and that it melts at the edges as the springtime of Mars comes round. In regard to its famous canals we are no nearer agreement. Some maintain that the markings are not really an objective feature; some hold that they are due to volcanic activity, and that similar markings are found on the moon; some believe that they are due to clouds; while Professor Lowell and others stoutly adhere to the familiar view that they are artificial canals, or the strips of vegetation along such canals. The question of the actual habitation of Mars is still open. We can say only that there is strong evidence of its possession of the conditions of life in some degree, and that living things, even on the earth, display a remarkable power of adaptation to widely differing conditions.

Passing over the 700 planetoids, which circulate between Mars and Jupiter, and for which we may account either by the absence of one large nucleus in that part of the nebulous stream or by the disturbing influence of Jupiter, we come to the largest planet of the system. Here we find a surprising confirmation of the theory of planetary development which we are following. Three hundred times heavier than the earth (or more than a trillion tons in weight), yet a thousand times less in volume than the sun, Jupiter ought, if our theory is correct, to be still red-hot. All the evidence conspires to suggest that it is. It has long been recognised that the shining disk of the planet is not a solid, but a cloud, surface. This impenetrable mass of cloud or vapour is drawn out in streams or belts from side to side, as the giant globe turns on its axis once in every ten hours. We cannot say if, or to what extent, these clouds consist of water-vapour. We can conclude only that this mantle of Jupiter is "a seething cauldron of vapours" (Lowell), and that, if the body beneath is solid, it must be very hot. A large red area, at one time 30,000 miles long, has more or less persisted on the surface for several decades, and it is generally interpreted, either as a red-hot surface, or as a vast volcanic vent, reflecting its glow upon the clouds. Indeed, the keen American observers, with their powerful telescopes, have detected a cherry-red glow on the edges of the cloud-belts across the disk; and more recent observation with the spectroscope seems to prove that Jupiter emits light from its surface analogous to that of the red stars. The conspicuous flattening of its poles is another feature that science would expect in a rapidly rotating liquid globe. In a word, Jupiter seems to be in the last stage of stellar development. Such, at some remote time, was our earth; such one day will be the sun.

The neighbouring planet Saturn supports the conclusion. Here again we have a gigantic globe, 28,000 miles in diameter, turning on its axis in the short space of ten hours; and here again we find the conspicuous flattening of the poles, the trailing belts of massed vapour across the disk, the red glow lighting the edges of the belts, and the spectroscopic evidence of an emission of light. Once more it is difficult to doubt that a highly heated body is wrapped in that thick mantle of vapour. With its ten moons and its marvellous ring-system—an enormous collection of fragments, which the influence of the planet or of its nearer satellites seems to have prevented from concentrating—Saturn has always been a beautiful object to observe; it is not less interesting in those features which we faintly detect in its disk.

The next planet, Uranus, 32,000 miles in diameter, seems to be another cloud-wrapt, greatly heated globe, if not, as some think, a sheer mass of vapours without a liquid core. Neptune is too dim and distant for profitable examination. It may be added, however, that the dense masses of gas which are found to surround the outer planets seem to confirm the nebular theory, which assumes that they were developed in the outer and lighter part of the material hurled from the sun.

From this encouraging survey of the sister-planets we return with more confidence to the story of the earth. I will not attempt to follow an imaginative scheme in regard to its early development. Take four photographs—one of a spiral nebula without knots in its arms, one of a nebula like that in Canes Venatici, one of the sun, and one of Jupiter—and you have an excellent illustration of the chief stages in its formation. In the first picture a section of the luminous arm of the nebula stretches thinly across millions of miles of space. In the next stage this material is largely collected in a luminous and hazy sphere, as we find in the nebula in Canes Venatici. The sun serves to illustrate a further stage in the condensation of this sphere. Jupiter represents a later chapter, in which the cooler vapours are wrapped close about the red-hot body of the planet. That seems to have been the early story of the earth. Some 6,000,000,000 billion tons of the nebulous matter were attracted to a common centre. As the particles pressed centreward, the temperature rose, and for a time the generation of heat was greater than its dissipation. Whether the earth ever shone as a small white star we cannot say. We must not hastily conclude that such a relatively small mass would behave like the far greater mass of a star, but we may, without attempting to determine its temperature, assume that it runs an analogous course.

One of the many features which I have indicated as pointing to a former fluidity of the earth may be explained here. We shall see in the course of this work that the mountain chains and other great irregularities of the earth's surface appear at a late stage in its development. Even as we find them to-day, they are seen to be merely slight ridges and furrows on the face of the globe, when we reflect on its enormous diameter, but there is good reason to think that in the beginning the earth was much nearer to a perfectly globular form. This points to a liquid or gaseous condition at one time, and the flattening of the sphere at the poles confirms the impression. We should hardly expect so perfect a rotundity in a body formed by the cool accretion of solid fragments and particles. It is just what we should expect in a fluid body, and the later irregularities of the surface are accounted for by the constant crumpling and wearing of its solid crust. Many would find a confirmation of this in the phenomena of volcanoes, geysers, and earthquakes, and the increase of the temperature as we descend the crust. But the interior condition of the earth, and the nature of these phenomena, are much disputed at present, and it is better not to rely on any theory of them. It is suggested that radium may be responsible for this subterraneous heat.

The next stage in the formation of the earth is necessarily one that we can reach only by conjecture. Over the globe of molten fire the vapours and gases would be suspended like a heavy canopy, as we find in Jupiter and Saturn to-day. When the period of maximum heat production was passed, however, the radiation into space would cause a lowering of the temperature, and a scum would form on the molten surface. As may be observed on the surface of any cooling vessel of fluid, the scum would stretch and crack; the skin would, so to say, prove too small for the body. The molten ocean below would surge through the crust, and bury it under floods of lava. Some hold that the slabs would sink in the ocean of metal, and thus the earth would first solidify in its deeper layers. There would, in any case, be an age-long struggle between the molten mass and the confining crust, until at length—to employ the old Roman conception of the activity of Etna—the giant was imprisoned below the heavy roof of rock.

Here again we seem to find evidence of the general correctness of the theory. The objection has been raised that the geologist does not find any rocks which he can identify as portions of the primitive crust of the earth. It seems to me that it would be too much to expect the survival at the surface of any part of the first scum that cooled on that fiery ocean. It is more natural to suppose that millions of years of volcanic activity on a prodigious scale would characterise this early stage, and the "primitive crust" would be buried in fragments, or dissolved again, under deep seas of lava. Now, this is precisely what we find, The oldest rocks known to the geologist—the Archaean rocks—are overwhelmingly volcanic, especially in their lower part. Their thickness, as we know them, is estimated at 50,000 feet; a thickness which must represent many millions of years. But we do not know how much thicker than this they may be. They underlie the oldest rocks that have ever been exposed to the gaze of the geologist. They include sedimentary deposits, showing the action of water, and even probable traces of organic remains, but they are, especially in their deeper and older sections, predominantly volcanic. They evince what we may call a volcanic age in the early story of the planet.

But before we pursue this part of the story further we must interpolate a remarkable event in the record—the birth of the moon. It is now generally believed, on a theory elaborated by Sir G. Darwin, that when the formation of the crust had reached a certain depth— something over thirty miles, it is calculated—it parted with a mass of matter, which became the moon. The size of our moon, in comparison with the earth, is so exceptional among the satellites which attend the planets of our solar system that it is assigned an exceptional origin. It is calculated that at that time the earth turned on its axis in the space of four or five hours, instead of twenty-four. We have already seen that the tidal influence of the sun has the effect of moderating the rotation of the planets. Now, this very rapid rotation of a liquid mass, with a thin crust, would (together with the instability occasioned by its cooling) cause it to bulge at the equator. The bulge would increase until the earth became a pear-shaped body. The small end of the pear would draw further and further away from the rest—as a drop of water does on the mouth of a tap—and at last the whole mass (some 5,000,000,000 cubic miles of matter) was broken off, and began to pursue an independent orbit round the earth.

There are astronomers who think that other cosmic bodies, besides our moon, may have been formed in this way. Possibly it is true of some of the double stars, but we will not return to that question. The further story of the moon, as it is known to astronomers, may be given in a few words. The rotational movement of the earth is becoming gradually slower on account of tidal influence; our day, in fact, becomes an hour longer every few million years. It can be shown that this had the effect of increasing the speed, and therefore enlarging the orbit, of the moon, as it revolved round the earth. As a result, the moon drew further and further away from the earth until it reached its present position, about 240,000 miles away. At the same time the tidal influence of the earth was lessening the rotational movement of the moon. This went on until it turned on its axis in the same period in which it revolves round the earth, and on this account it always presents the same face to the earth.

Through what chapters of life the moon may have passed in the meantime it is impossible to say. Its relatively small mass may have been unable to keep the lighter gases at its surface, or its air and water may, as some think, have been absorbed. It is to-day practically an airless and waterless desert, alternating between the heat of its long day and the intense cold of its long night. Careful observers, such as Professor Pickering, think that it may still have a shallow layer of heavy gases at its surface, and that this may permit the growth of some stunted vegetation during the day. Certain changes of colour, which are observed on its surface, have been interpreted in that sense. We can hardly conceive any other kind of life on it. In the dark even the gases will freeze on its surface, as there is no atmosphere to retain the heat. Indeed, some students of the moon (Fauth, etc.) believe that it is an unchanging desert of ice, bombarded by the projectiles of space.

An ingenious speculation as to the effect on the earth of this dislodgment of 5,000,000,000 cubic miles of its substance is worth noting. It supposes that the bed of the Pacific Ocean represents the enormous gap torn in its side by the delivery of the moon. At each side of this chasm the two continents, the Old World and the New, would be left floating on their molten ocean; and some have even seen a confirmation of this in the lines of crustal weakness which we trace, by volcanoes and earthquakes, on either side of the Pacific. Others, again, connect the shape of our great masses of land, which generally run to a southern point, with this early catastrophe. But these interesting speculations have a very slender basis, and we will return to the story of the development of the earth.

The last phase in preparation for the appearance of life would be the formation of the ocean. On the lines of the generally received nebular hypothesis this can easily be imagined, in broad outline. The gases would form the outer shell of the forming planet, since the heavier particles would travel inward. In this mixed mass of gas the oxygen and hydrogen would combine, at a fitting temperature, and form water. For ages the molten crust would hold this water suspended aloft as a surrounding shell of cloud, but when the surface cooled to about 380 degrees C. (Sollas), the liquid would begin to pour on it. A period of conflict would ensue, the still heated crust and the frequent volcanic outpours sending the water back in hissing steam to the clouds. At length, and now more rapidly, the temperature of the crust would sink still lower, and a heated ocean would settle upon it, filling the hollows of its irregular surface, and washing the bases of its outstanding ridges. From that time begins the age-long battle of the land and the water which, we shall see, has had a profound influence on the development of life.

In deference to the opinion of a number of geologists we must glance once more at the alternative view of the planetesimal school. In their opinion the molecules of water were partly attracted to the surface out of the disrupted matter, and partly collected within the porous outer layers of the globe. As the latter quantity grew, it would ooze upwards, fill the smaller depressions in the crust, and at length, with the addition of the attracted water, spread over the irregular surface. There is an even more important difference of opinion in regard to the formation of the atmosphere, but we may defer this until the question of climate interests us. We have now made our globe, and will pass on to that early chapter of its story in which living things make their appearance.

To some it will seem that we ought not to pass from the question of origin without a word on the subject of the age of the earth. All that one can do, however, is to give a number of very divergent estimates. Physicists have tried to calculate the age of the sun from the rate of its dissipation of heat, and have assigned, at the most, a hundred million years to our solar system; but the recent discovery of a source of heat in the disintegration of such metals as radium has made their calculations useless. Geologists have endeavoured, from observation of the action of geological agencies to-day, to estimate how long it will have taken them to form the stratified crust of the earth; but even the best estimates vary between twenty-five and a hundred million years, and we have reason to think that the intensity of these geological agencies may have varied in different ages. Chemists have calculated how long it would take the ocean, which was originally fresh water, to take up from the rocks and rivers the salt which it contains to-day; Professor Joly has on this ground assigned a hundred million years since the waters first descended upon the crust. We must be content to know that the best recent estimates, based on positive data, vary between fifty and a hundred million years for the story which we are now about to narrate. The earlier or astronomical period remains quite incalculable. Sir G. Darwin thinks that it was probably at least a thousand million years since the moon was separated from the earth. Whatever the period of time may be since some cosmic cataclysm scattered the material of our solar system in the form of a nebula, it is only a fraction of that larger and illimitable time which the evolution of the stars dimly suggests to the scientific imagination.

THE GEOLOGICAL SERIES

[The scale of years adopted—50,000,000 for the stratified rocks—is merely an intermediate between conflicting estimates.]

ERA.	PERIOD.	RELATIVE LENGTH.
Quaternary years	Holocene	500,000
	Pleistocene	
Tertiary years or	Pliocene	5,500,000
	Miocene	

Cenozoic	Oligocene	
	Eocene	
Secondary	Cretaceous	7,200,000
years		
or	Jurassic	3,600,000
"		
Mesozoic	Triassic	2,500,000
"		
Primary	Permian	2,800,000
years		
or	Carboniferous	6,200,000
"		
Palaeozoic	Devonian	8,000,000
"		
	Silurian	5,400,000
"		
	Ordovician	5,400,000
"		
	Cambrian	8,000,000
"		
Archaean	Keweenawan	Unknown
(probably		
	Animikie	at least
	Huronian	50,000,000
years)		
	Keewatin	
	Laurentian	

CHAPTER V. THE BEGINNING OF LIFE

There is, perhaps, no other chapter in the chronicle of the earth that we approach with so lively an interest as the chapter which should record the first appearance of life. Unfortunately, as far as the authentic memorials of the past go, no other chapter is so impenetrably obscure as this. The reason is simple. It is a familiar saying that life has written its own record, the long-drawn record of its dynasties and its deaths, in the rocks. But there were millions of years during which life had not yet learned to write its record, and further millions of years the record of which has been irremediably destroyed. The first volume of the geological chronicle of the earth is the mass of the Archaean (or "primitive") rocks. What the actual magnitude of that volume, and the span of time it covers, may be, no geologist can say. The Archaean rocks still solidly underlie the lowest depth he has ever reached. It is computed, however, that these rocks, as far as they are known to us, have a total depth of nearly ten miles, and seem therefore to represent at least half the story of the earth from the time when it rounded into a globe, or cooled sufficiently to endure the presence of oceans.

Yet all that we read of the earth's story during those many millions of years could be told in a page or two. That section of geology is still in its infancy, it is true. A day may come when science will decipher a long and instructive narrative in the masses of quartz and gneiss, and the layers of various kinds, which it calls the Archaean rocks. But we may say with confidence that it will not discover in them more than a few stray syllables of the earlier part, and none whatever of the earliest part, of the epic of living nature. A few fossilised remains of somewhat advanced organisms, such as shell-fish and worms, are found in the higher and later rocks of the series, and more of the same comparatively high types will probably appear. In the earlier strata, representing an earlier stage of life, we

66

find only thick seams of black shale, limestone, and ironstone, in which we seem to see the ashes of primitive organisms, cremated in the appalling fires of the volcanic age, or crushed out of recognition by the superimposed masses. Even if some wizardry of science were ever to restore the forms that have been reduced to ashes in this Archaean crematorium, it would be found that they are more or less advanced forms, far above the original level of life. No trace will ever be found in the rocks of the first few million years in the calendar of life.

The word impossible or unknowable is not lightly uttered in science to-day, but there is a very plain reason for admitting it here. The earliest living things were at least as primitive of nature as the lowest animals and plants we know to-day, and these, up to a fair level of organisation, are so soft of texture that, when they die, they leave no remains which may one day be turned into fossils. Some of them, indeed, form tiny shells of flint or lime, or, like the corals, make for themselves a solid bed; but this is a relatively late and higher stage of development. Many thousands of species of animals and plants lie below that level. We are therefore forced to conclude, from the aspect of living nature to-day, that for ages the early organisms had no hard and preservable parts. In thus declaring the impotence of geology, however, we are at the same time introducing another science, biology, which can throw appreciable light on the evolution of life. Let us first see what geology tells us about the infancy of the earth.

The distribution of the early rocks suggests that there was comparatively little dry land showing above the surface of the Archaean ocean. Our knowledge of these rocks is not at all complete, and we must remember that some of this primitive land may be now under the sea or buried in unsuspected regions. It is significant, however, that, up to the present, exploration seems to show that in those remote ages only about one-fifth of our actual land-surface stood above the level of the waters. Apart from a patch of some 20,000 square miles of what is now Australia, and smaller patches in Tasmania, New Zealand, and India, nearly the whole of this land was in the far North. A considerable area of eastern Canada had emerged, with lesser islands standing out to the west and south of North America. Another large area lay round the basin of the Baltic; and as Greenland, the Hebrides, and the extreme tip of Scotland, belong to the same age, it is believed that a continent, of which they are fragments, united America and Europe across the North Atlantic. Of the rest of what is now Europe there were merely large islands— one on the border of England and Wales, others in France, Spain, and Southern Germany. Asia was represented by a large area in China and Siberia, and an island or islands on the site of India. Very little of Africa or South America existed.

It will be seen at a glance that the physical story of the earth from that time is a record of the emergence from the waters of larger continents and the formation of lofty chains of mountains. Now this world-old battle of land and sea has been waged with varying fortune from age to age, and it has been one of the most important factors in the development of life. We are just beginning to realise what a wonderful light it throws on the upward advance of animals and plants. No one in the scientific world to-day questions that, however imperfect the record may be, there has been a continuous development of life from the lowest level to the highest. But why there was advance at all, why the primitive microbe climbs the scale of being, during millions of years, until it reaches the stature of humanity, seems to many a profound mystery. The solution of this mystery begins to break upon us when we contemplate, in the geological record, the prolonged series of changes in the face of the earth itself, and try to realise how these changes must have impelled living things to fresh and higher adaptations to their changing surroundings.

Imagine some early continent with its population of animals and plants. Each bay, estuary, river, and lake, each forest and marsh and solid plain, has its distinctive inhabitants. Imagine this continent slowly sinking into the sea, until the advancing arms of the salt water meet across it, mingling their diverse populations in a common world, making the fresh-water lake brackish or salt, turning the dry land into swamp, and flooding the forest. Or suppose, on the other hand, that the land rises, the marsh is drained, the genial climate succeeded by an icy cold, the luscious vegetation destroyed, the whole animal population compelled to change its habits and its food. But this is no imaginary picture. It is the actual story of the earth during millions of years, and it is chiefly in the light of these vast and exacting changes in the environment that we are going to survey the panorama of the advance of terrestrial life.

For the moment it will be enough to state two leading principles. The first is that there is no such thing as a "law of evolution" in the sense in which many people understand that phrase. It is now sufficiently well known that, when science speaks of a law, it does not mean that there is some rule that things MUST act in such and such a way. The law is a mere general expression of the fact that they DO act in that way. But many imagine that there is some principle within the living organism which impels it onward to a higher level of organisation. That is entirely an error. There is no "law of progress." If an animal is fitted to secure its livelihood and breed posterity in certain surroundings, it may remain unchanged indefinitely if these surroundings do not materially change. So the duckmole of Australia and the tuatara of New Zealand have retained primitive features for millions of years; so the aboriginal Australian and the Fuegian have remained stagnant, in their isolation, for a hundred thousand years or more; so the Chinaman, in his geographical isolation, has remained unchanged for two thousand years. There is no more a "conservative instinct" in Chinese than there is a "progressive instinct" in Europeans. The difference is one of history and geography, as we shall see.

To make this important principle still clearer, let us imagine some primitive philosopher observing the advance of the tide over a level beach. He must discover two things: why the water comes onward at all, and why it advances along those particular channels. We shall see later how men of science explain or interpret the mechanism in a living thing which enables it to advance, when it does advance. For the present it is enough to say that new-born animals and plants are always tending to differ somewhat from their parents, and we now know, by experiment, that when some exceptional influence is brought to bear on the parent, the young may differ considerably from her. But, if the parents were already in harmony with their environment, these variations on the part of the young are of no consequence. Let the environment alter, however, and some of these variations may chance to make the young better fitted than the parent was. The young which happen to have the useful variation will have an advantage over their brothers or sisters, and be more likely to survive and breed the next generation. If the change in the environment (in the food or climate, for instance) is prolonged and increased for hundreds of thousands of years, we shall expect to find a corresponding change in the animals and plants.

We shall find such changes occurring throughout the story of the earth. At one important point in the story we shall find so grave a revolution in the face of nature that twenty-nine out of every thirty species of animals and plants on the earth are annihilated. Less destructive and extreme changes have been taking place during nearly the whole of the period we have to cover, entailing a more gradual alteration of the structure of animals and plants; but we shall repeatedly find them culminating in very great changes of climate, or of the distribution of land and water, which have subjected the living population of the earth to the most searching tests and promoted every variation toward a more effective organisation. [*]

And the second guiding principle I wish to lay down in advance is that these great changes in the face of the earth, which explain the progress of organisms, may very largely be reduced to one simple agency—the battle of the land and the sea. When you gaze at some line of cliffs that is being eaten away by the waves, or reflect on the material carried out to sea by the flooded river, you are—paradoxical as it may seem—beholding a material process that has had a profound influence on the development of life. The Archaean continent that we described was being reduced constantly by the wash of rain, the scouring of rivers, and the fretting of the waves on the coast. It is generally thought that these wearing agencies were more violent in early times, but that is disputed, and we will not build on it. In any case, in the course of time millions of tons of matter were scraped off the Archaean continent and laid on the floor of the sea by its rivers. This meant a very serious alteration of pressure or weight on the surface of the globe, and was bound to entail a reaction or restoration of the balance.

The rise of the land and formation of mountains used to be ascribed mainly to the cooling and shrinking of the globe of the earth. The skin (crust), it was thought, would become too large for the globe as it shrank, and would wrinkle outwards, or pucker up into mountain-chains. The position of our greater mountain-chains sprawling across half the earth (the Pyrenees to the Himalaya, and the Rocky Mountains to the Andes), seems to confirm this, but the question of the interior of the earth is obscure and disputed, and geologists generally conceive the rise of land and formation of mountains in a different way. They are due probably to the alteration of pressure on the crust in combination with the instability of the interior. The floors of the seas would sink still lower under their colossal burdens, and this would cause some draining of the land-surface. At the same time the heavy pressure below the seas and the lessening of pressure over the land would provoke a reaction. Enormous masses of rock would be forced toward and underneath the land-surface, bending, crumpling, and upheaving it as if its crust were but a leather coat. As a result, masses of land would slowly rise above the plain, to be shaped into hills and valleys by the hand of later time, and fresh surfaces would be dragged out of the deep, enlarging the fringes of the primitive continents, to be warped and crumpled in their turn at the next era of pressure.

In point of geological fact, the story of the earth has been one prolonged series of changes in the level of land and water, and in their respective limits. These changes have usually been very gradual, but they have always entailed changes (in climate, etc.) of the greatest significance in the evolution of life. What was the swampy soil of England in the Carboniferous period is now sometimes thousands of feet beneath us; and what was the floor of a deep ocean over much of Europe and Asia at another time is now to be found on the slopes of lofty Alps, or 20,000 feet above the sea-level in Thibet. Our story of terrestrial life will be, to a great extent, the story of how animals and plants changed their structure in the long series of changes which this endless battle of land and sea brought over the face of the earth.

As we have no recognisable remains of the animals and plants of the earliest age, we will not linger over the Archaean rocks. Starting from deep and obscure masses of volcanic matter, the geologist, as he travels up the series of Archaean rocks, can trace only a dim and most unsatisfactory picture of those remote times. Between outpours of volcanic floods he finds, after a time, traces that an ocean and rivers are wearing away the land. He finds seams of carbon among the rocks of the second division of the Archaean (the Keewatin), and deduces from this that a dense sea-weed population already covered the floor of the ocean. In the next division (the Huronian) he finds the traces of extensive ice-action strangely lying between masses of volcanic rock, and sees that thousands of square miles of eastern North America were then covered with an ice-sheet. Then fresh floods of molten matter are poured out from the depths below; then the sea floods the land for a time; and at last it makes its final emergence as the first definitive part of the North American continent, to enlarge, by successive fringes, to the continent of to-day. [*]

    * I am quoting Professor Coleman's summary of Archaean
    research in North America (Address to the Geological Section
    of the British Association, 1909). Europe, as a continent,
    has had more "ups and downs" than America in the course of
    geological time.

This meagre picture of the battle of land and sea, with interludes of great volcanic activity and even of an ice age, represents nearly all we know of the first half of the world's story from geology. It is especially disappointing in regard to the living population. The very few fossils we find in the upper Archaean rocks are so similar to those we shall discuss in the next chapter that we may disregard them, and the seams of carbon-shales, iron-ore, and limestone, suggest only, at the most, that life was already abundant. We must turn elsewhere for some information on the origin and early development of life.

The question of the origin of life I will dismiss with a brief account of the various speculations of recent students of science. Broadly speaking, their views fall into three classes. Some think that the germs of life may have come to the earth from some other body in the universe; some think that life was evolved out of non-living matter in the early ages of the earth, under exceptional conditions which we do not at present know, or can only dimly conjecture; and some think that life is being evolved from non-life in nature to-day, and always has been so evolving. The majority of scientific men merely assume that the earliest living things were no exception to the general process of evolution, but think that we have too little positive knowledge to speculate profitably on the manner of their origin.

The first view, that the germs of life may have come to this planet on a meteoric visitor from some other world, as a storm-driven bird may take its parasites to some distant island, is not without adherents to-day. It was put forward long ago by Lord Kelvin and others; it has been revived by the distinguished Swede, Professor Svante Arrhenius. The scientific objection to it is that the more intense (ultra-violet) rays of the sun would frill such germs as they pass through space. But a broader objection, and one that may dispense us from dwelling on it, is that we gain nothing by throwing our problems upon another planet. We have no ground for supposing that the earth is less capable of evolving life than other planets.

The second view is that, when the earth had passed through its white-hot stage, great masses of very complex chemicals, produced by the great heat, were found on its surface. There is one complex chemical substance in particular, called cyanogen, which is either an important constituent of living matter, or closely akin to it. Now we need intense heat to produce this substance in the laboratory. May we not suppose that masses of it were produced during the incandescence of the earth, and that, when the waters descended, they passed through a series of changes which culminated in living plasm? Such is the "cyanogen hypothesis" of the origin of life, advocated by able physiologists such as Pfluger, Verworn, and others. It has the merit of suggesting a reason why life may not be evolving from non-life in nature to-day, although it may have so evolved in the Archaean period.

Other students suggest other combinations of carbon-compounds and water in the early days. Some suggest that electric action was probably far more intense in those ages; others think that quantities of radium may have been left at the surface. But the most important of these speculations on the origin of life in early times, and one that has the merit of not assuming any essentially different conditions then than we find now, is contained in a recent pronouncement of one of the greatest organic chemists in Europe, Professor Armstrong. He says that such great progress has been made in his science—the science of the chemical processes in living things—that "their cryptic character seems to have disappeared almost suddenly." On the strength of this new knowledge of living matter, he ventures to say that "a series of lucky accidents" could account for the first formation of living things out of non-living matter in Archaean times. Indeed, he goes further. He names certain inorganic substances, and says that the blowing of these into pools by the wind on the primitive planet would set afoot chemical combinations which would issue in the production of living matter. [*]

* See his address in Nature, vol. 76, p. 651. For other

speculations see Verworn's "General Physiology," Butler
    Burke's "Origin of Life" (1906), and Dr. Bastian's "Origin
    of Life" (1911).

It is evident that the popular notion that scientific men have declared that life cannot be evolved from non-life is very far astray. This blunder is usually due to a misunderstanding of the dogmatic statement which one often reads in scientific works that "every living thing comes from a living thing." This principle has no reference to remote ages, when the conditions may have been different. It means that to-day, within our experience, the living thing is always born of a living parent. However, even this is questioned by some scientific men of eminence, and we come to the third view.

Professor Nageli, a distinguished botanist, and Professor Haeckel, maintain that our experience, as well as the range of our microscopes, is too limited to justify the current axiom. They believe that life may be evolving constantly from inorganic matter. Professor J. A. Thomson also warns us that our experience is very limited, and, for all we know, protoplasm may be forming naturally in our own time. Mr. Butler Burke has, under the action of radium, caused the birth of certain minute specks which strangely imitate the behaviour of bacteria. Dr. Bastian has maintained for years that he has produced living things from non-living matter. In his latest experiments, described in the book quoted, purely inorganic matter is used, and it is previously subjected, in hermetically sealed tubes, to a heat greater than what has been found necessary to kill any germs whatever.

Evidently the problem of the origin of life is not hopeless, but our knowledge of the nature of living matter is still so imperfect that we may leave detailed speculation on its origin to a future generation. Organic chemistry is making such strides that the day may not be far distant when living matter will be made by the chemist, and the secret of its origin revealed. For the present we must be content to choose the more plausible of the best-informed speculations on the subject.

But while the origin of life is obscure, the early stages of its evolution come fairly within the range of our knowledge. To the inexpert it must seem strange that, whereas we must rely on pure speculation in attempting to trace the origin of life, we can speak with more confidence of those early developments of plants and animals which are equally buried in the mists of the Archaean period. Have we not said that nothing remains of the procession of organisms during half the earth's story but a shapeless seam of carbon or limestone?

A simple illustration will serve to justify the procedure we are about to adopt. Suppose that the whole of our literary and pictorial references to earlier stages in the development of the bicycle, the locomotive, or the loom, were destroyed. We should still be able to retrace the phases of their evolution, because we should discover specimens belonging to those early phases lingering in our museums, in backward regions, and elsewhere. They might yet be useful in certain environments into which the higher machines have not penetrated. In the same way, if all the remains of prehistoric man and early civilisation were lost, we could still fairly retrace the steps of the human race, by gathering the lower tribes and races, and arranging them in the order of their advancement. They are so many surviving illustrations of the stages through which mankind as a whole has passed.

Just in the same way we may marshal the countless species of animals and plants to-day in such order that they will, in a general way, exhibit to us the age-long procession of life. From the very start of living evolution certain forms dropped out of the onward march, and have remained, to our great instruction, what their ancestors were millions of years ago. People create a difficulty for themselves by imagining that, if evolution is true, all animals must evolve. A glance at our own fellows will show the error of this. Of one family of human beings, as a French writer has said, one only becomes a Napoleon; the others remain Lucien, Jerome, or Joseph. Of one family of animals or trees, some advance in one or other direction; some remain at the original level. There is no "law of progress." The accidents of the world and hereditary endowment impel some onward, and do not impel others. Hence at nearly every great stage in the upward procession through the ages some regiment of plants or animals has dropped out, and it represents to-day the stage of life at which it ceased to progress. In other words, when we survey the line of the hundreds of thousands of species which we find in nature to-day, we can trace, amid their countless variations and branches, the line of organic evolution in the past; just as we could, from actual instances, study the evolution of a British house, from the prehistoric remains in Devonshire to a mansion in Park Lane or a provincial castle.

Another method of retracing the lost early chapters in the development of life is furnished by embryology. The value of this method is not recognised by all embryologists, but there are now few authorities who question the substantial correctness of it, and we shall, as we proceed, see some remarkable applications of it. In brief, it is generally admitted that an animal or plant is apt to reproduce, during its embryonic development, some of the stages of its ancestry in past time. This does not mean that a higher animal, whose ancestors were at one time worms, at another time fishes, and at a later time reptiles, will successively take the form of a little worm, a little fish, and a little reptile. The embryonic life itself has been subject to evolution, and this reproduction of ancestral forms has been proportionately disturbed. Still, we shall find that animals will tend, in their embryonic development, to reproduce various structural features which can only be understood as reminiscences of ancestral organs. In the lower animals the reproduction is much less disturbed than in the higher, but even in the case of man this law is most strikingly verified. We shall find it useful sometimes at least in confirming our conclusions as to the ancestry of a particular group.

We have, therefore, two important clues to the missing chapters in the story of evolution. Just as the scheme of the evolution of worlds is written broadly across the face of the heavens to-day, so the scheme of the evolution of life is written on the face of living nature; and it is written again, in blurred and broken characters, in the embryonic development of each individual. With these aids we set out to restore the lost beginning of the epic of organic evolution.

CHAPTER VI. THE INFANCY OF THE EARTH

The long Archaean period, into which half the story of the earth is so unsatisfactorily packed, came to a close with a considerable uplift of the land. We have seen that the earth at times reaches critical stages owing to the transfer of millions of tons of matter from the land to the depths of the ocean, and the need to readjust the pressure on the crust. Apparently this stage is reached at the end of the Archaean, and a great rise of the land—probably protracted during hundreds of thousands of years—takes place. The shore-bottoms round the primitive continent are raised above the water, their rocks crumpling like plates of lead under the overpowering pressure. The sea retires with its inhabitants, mingling their various provinces, transforming their settled homes. A larger continent spans the northern ocean of the earth.

In the shore-waters of this early continent are myriads of living things, representing all the great families of the animal world below the level of the fish and the insect. The mud and sand in which their frames are entombed, as they die, will one day be the "Cambrian" rocks of the geologist, and reveal to him their forms and suggest their habits. No great volcanic age will reduce them to streaks of shapeless carbon. The earth now buries its dead, and from their petrified remains we conjure up a picture of the swarming life of the Cambrian ocean.

A strange, sluggish population burrows in the mud, crawls over the sand, adheres to the rocks, and swims among the thickets of sea-weed. The strangest and most formidable, though still too puny a thing to survive in a more strenuous age, is the familiar Trilobite of the geological museum; a flattish animal with broad, round head, like a shovel, its back covered with a three-lobed shell, and a number of fine legs or swimmers below. It burrows in the loose bottom, or lies in it with its large compound eyes peeping out in search of prey. It is the chief representative of the hard-cased group (Crustacea) which will later replace it with the lobster, the shrimp, the crab, and the water-flea. Its remains form from a third to a fourth of all the buried Cambrian skeletons. With it, swimming in the water, are smaller members of the same family, which come nearer to our familiar small Crustacea.

Shell-fish are the next most conspicuous inhabitants. Molluscs are already well represented, but the more numerous are the more elementary Brachiopods ("lampshells"), which come next to the Trilobites in number and variety. Worms (or Annelids) wind in and out of the mud, leaving their tracks and tubes for later ages. Strange ball or cup-shaped little animals, with a hard frame, mounted on stony stalks and waving irregular arms to draw in the food-bearing water, are the earliest representatives of the Echinoderms. Some of these Cystids will presently blossom into the wonderful sea-lily population of the next age, some are already quitting their stalks, to become the free-moving star-fish, of which a primitive specimen has been found in the later Cambrian. Large jelly-fishes (of which casts are preserved) swim in the water; coral-animals lay their rocky foundations, but do not as yet form reefs; coarse sponges rise from the floor; and myriads of tiny Radiolaria and Thalamophores, with shells of flint and lime, float at the surface or at various depths.

This slight sketch of the Cambrian population shows us that living things had already reached a high level of development. Their story evidently goes back, for millions of years, deep into those mists of the Archaean age which we were unable to penetrate. We turn therefore to the zoologist to learn what he can tell us of the origin and family-relations of these Cambrian animals, and will afterwards see how they are climbing to higher levels under the eye of the geologist.

At the basis of the living world of to-day is a vast population of minute, generally microscopic, animals and plants, which are popularly known as "microbes." Each consists, in scientific language, of one cell. It is now well known that the bodies of the larger animals and plants are made up of millions of these units of living matter, or cells—the atoms of the organic world—and I need not enlarge on it. But even a single cell lends itself to infinite variety of shape, and we have to penetrate to the very lowest level of this luxuriant world of one-celled organisms to obtain some idea of the most primitive living things. Properly speaking, there were no "first living things." It cannot be doubted by any student of nature that the microbe developed so gradually that it is as impossible to fix a precise term for the beginning of life as it is to say when the night ends and the day begins. In the course of time little one-celled living units appeared in the waters of the earth, whether in the shallow shore waters or on the surface of the deep is a matter of conjecture.

We are justified in concluding that they were at least as rudimentary in structure and life as the lowest inhabitants of nature to-day. The distinction of being the lowest known living organisms should, I think, be awarded to certain one-celled vegetal organisms which are very common in nature. Minute simple specks of living matter, sometimes less than the five-thousandth of an inch in diameter, these lowly Algae are so numerous that it is they, in their millions, which cover moist surfaces with the familiar greenish or bluish coat. They have no visible organisation, though, naturally, they must have some kind of structure below the range of the microscope. Their life consists in the absorption of food-particles, at any point of their surface, and in dividing into two living microbes, instead of dying, when their bulk increases. A very lowly branch of the Bacteria (Nitrobacteria) sometimes dispute their claim to the lowest position in the hierarchy of living nature, but there is reason to suspect that these Bacteria may have degenerated from a higher level.

Here we have a convenient starting-point for the story of life, and may now trace the general lines of upward development. The first great principle to be recognised is the early division of these primitive organisms into two great classes, the moving and the stationary. The clue to this important divergence is found in diet. With exceptions on both sides, we find that the non-moving microbes generally feed on inorganic matter, which they convert into plasm; the moving microbes generally feed on ready-made plasm—on the living non-movers, on each other, or on particles of dead organic matter. Now, inorganic food is generally diffused in the waters, so that the vegetal feeders have no incentive to develop mobility. On the other hand, the power to move in search of their food, which is not equally diffused, becomes a most important advantage to the feeders on other organisms. They therefore develop various means of locomotion. Some flow or roll slowly along like tiny drops of oil on an inclined surface; others develop minute outgrowths of their substance, like fine hairs, which beat the water as oars do. Some of them have one strong oar, like the gondolier (but in front of the boat); others have two or more oars; while some have their little flanks bristling with fine lashes, like the flanks of a Roman galley.

If we imagine this simple principle at work for ages among the primitive microbes, we understand the first great division of the living world, into plants and animals. There must have been a long series of earlier stages below the plant and animal. In fact, some writers insist that the first organisms were animal in nature, feeding on the more elementary stages of living matter. At last one type develops chlorophyll (the green matter in leaves), and is able to build up plasm out of inorganic matter; another type develops mobility, and becomes a parasite on the plant world. There is no rigid distinction of the two worlds. Many microscopic plants move about just as animals do, and many animals live on fixed stalks; while many plants feed on organic matter. There is so little "difference of nature" between the plant and the animal that the experts differ in classifying some of these minute creatures. In fact, we shall often find plants and animals crossing the line of division. We shall find animals rooting themselves to the floor, like plants, though they will generally develop arms or streamers for bringing the food to them; and we shall find plants becoming insect-catchers. All this merely shows that the difference is a natural tendency, which special circumstances may overrule. It remains true that the great division of the organic world is due to a simple principle of development; difference of diet leads to difference of mobility.

But this simple principle will have further consequences of a most important character. It will lead to the development of mind in one half of living nature and leave it undeveloped in the other. Mind, as we know it in the lower levels of life, is not confined to the animal at all. Many even of the higher plants are very delicately sensitive to stimulation, and at the lowest level many plants behave just like animals. In other words, this sensitiveness to stimuli, which is the first form of mind, is distributed according to mobility. To the motionless organism it is no advantage; to the pursuing and pursued organism it is an immense advantage, and is one of the chief qualities for natural selection to foster.

For the moment, however, we must glance at the operation of this and other natural principles in the evolution of the one-celled animals and plants, which we take to represent the primitive population of the earth. As there are tens of thousands of different species even of "microbes," it is clear that we must deal with them in a very summary way. The evolution of the plant I reserve for a later chapter, and I must be content to suggest the development of one-celled animals on very broad lines. When some of the primitive cells began to feed on each other, and develop mobility, it is probable that at least two distinct types were evolved, corresponding to the two lowest animal organisms in nature to-day. One of these is a very minute and very common (in vases of decaying flowers, for instance) speck of plasm, which moves about by lashing the water with a single oar (flagellum), or hair-like extension of its substance. This type, however, which is known as the Flagellate, may be derived from the next, which we will take as the primitive and fundamental animal type. It is best seen in the common and familiar Amoeba, a minute sac of liquid or viscid plasm, often not more than a hundredth of an inch in diameter. As its "skin" is merely a finer kind of the viscous plasm, not an impenetrable membrane, it takes in food at any part of its surface, makes little "stomachs," or temporary cavities, round the food at any part of its interior, ejects the useless matter at any point, and thrusts out any part of its body as temporary "arms" or "feet."

Now it is plain that in an age of increasing microbic cannibalism the toughening of the skin would be one of the first advantages to secure survival, and this is, in point of fact, almost the second leading principle in early development. Naturally, as the skin becomes firmer, the animal can no longer, like the Amoeba, take food at, or make limbs of, any part of it. There must be permanent pores in the membrane to receive food or let out rays of the living substance to act as oars or arms. Thus we get an immense variety amongst these Protozoa, as the one-celled animals are called. Some (the Flagellates) have one or two stout oars; some (the Ciliates) have numbers of fine hairs (or cilia). Some have a definite mouth-funnel, but no stomach, and cilia drawing the water into it. Some (Vorticella, etc.), shrinking from the open battlefield, return to the plant-principle, live on stalks, and have wreaths of cilia round the open mouth drawing the water to them. Some (the Heliozoa) remain almost motionless, shooting out sticky rays of their matter on every side to catch the food. Some form tubes to live in; some (Coleps) develop horny plates for armour; and others develop projectiles to pierce their prey (stinging threads).

This miniature world is full of evolutionary interest, but it is too vast for detailed study here. We will take one group, which we know to have been already developed in the Cambrian, and let a study of its development stand for all. In every lecture or book on "the beauties of the microscope" we find, and are generally greatly puzzled by, minute shells of remarkable grace and beauty that are formed by some of these very elementary animals They are the Radiolaria (with flinty shells, as a rule) and the Thalamophora (with chalk frames). Evolution furnishes a simple key to their remarkable structure.

As we saw, one of the early requirements to be fostered by natural selection in the Archaean struggle for life was a "thick skin," and the thick skin had to be porous to let the animal shoot out its viscid substance in rays and earn its living. This stage above the Amoeba is beautifully illustrated in the sun-animalcules (Heliozoa). Now the lowest types of Radiolaria are of this character. They have no shell or framework at all. The next stage is for the little animal to develop fine irregular threads of flint in its skin, a much better security against the animal-eater. These animalcules, it must be recollected, are bits of almost pure plasm, and, as they live in crowds, dividing and subdividing, but never dying, make excellent mouthfuls for a small feeder. Those with the more flint in their skins were the more apt to survive and "breed." The threads of flint increase until they form a sort of thorn-thicket round a little social group, or a complete lattice round an individual body. Next, spikes or spines jut out from the lattice, partly for additional protection, partly to keep the little body afloat at the surface of the sea. In this way we get a bewildering variety and increasing complexity of forms, ascending in four divergent lines from the naked ancestral type to the extreme grace and intricacy of the Calocyclas monumentum or the Lychnaspis miranda. These, however, are rare specimens in the 4000 species of Radiolaria. I have hundreds of them, on microscopic slides, which have no beauty and little regularity of form. We see a gradual evolution, on utilitarian principles, as we run over the thousands of forms; and, when we recollect the inconceivable numbers in which these little animals have lived and struggled for life—passively—during tens of millions of years, we are not surprised at the elaborate protective frames of the higher types.

The Thalamophores, the sister-group of one-celled animals which largely compose our chalk and much of our limestone, are developed on the same principle. The earlier forms seem to have lived in a part of the ocean where silica was scarce, and they absorbed and built their protective frames of lime. In the simpler types the frame is not unlike a wide-necked bottle, turned upside-down. In later forms it takes the shape of a spirally coiled series of chambers, sometimes amounting to several thousand. These wonderful little houses are not difficult to understand. The original tiny animal covers itself with a coat of lime. It feeds, grows, and bulges out of its chamber. The new part of its flesh must have a fresh coat, and the process goes on until scores, or hundreds, or even thousands, of these tiny chambers make up the spiral shell of the morsel of living matter.

With this brief indication of the mechanical principles which have directed the evolution of two of the most remarkable groups of the one-celled animals we must be content, or the dimensions of this volume will not enable us even to reach the higher and more interesting types. We must advance at once to the larger animals, whose bodies are composed of myriads of cells.

The social tendency which pervades the animal world, and the evident use of that tendency, prepare us to understand that the primitive microbes would naturally come in time to live in clusters. Union means effectiveness in many ways, even when it does not mean strength. We have still many loose associations of one-celled animals in nature, illustrating the approach to a community life. Numbers of the Protozoa are social; they live either in a common jelly-like matrix, or on a common stalk. In fact, we have a singularly instructive illustration of the process in the evolution of the sponges.

It is well known that the horny texture to which we commonly give the name of sponge is the former tenement and shelter of a colony of one-celled animals, which are the real Sponges. In other groups the structure is of lime; in others, again, of flinty material. Now, the Sponges, as we have them to-day, are so varied, and start from so low a level, that no other group of animals "illustrates so strikingly the theory of evolution," as Professor Minchin says. We begin with colonies in which the individuals are (as in Proterospongia) irregularly distributed in their jelly-like common bed, each animal lashing the water, as stalked Flagellates do, and bringing the food to it. Such a colony would be admirable food for an early carnivore, and we soon find the protective principle making it less pleasant for the devourer. The first stage may be—at least there are such Sponges even now—

that the common bed is strewn or sown with the cast shells of Radiolaria. However that may be, the Sponges soon begin to absorb the silica or lime of the sea-water, and deposit it in needles or fragments in their bed. The deposit goes on until at last an elaborate framework of thorny, or limy, or flinty material is constructed by the one-celled citizens. In the higher types a system of pores or canals lets the food-bearing water pass through, as the animals draw it in with their lashes; in the highest types the animals come still closer together, lining the walls of little chambers in the interior.

Here we have a very clear evolutionary transition from the solitary microbe to a higher level, but, unfortunately, it does not take us far. The Sponges are a side-issue, or cul de sac, from the Protozoic world, and do not lead on to the higher. Each one-celled unit remains an animal; it is a colony of unicellulars, not a many-celled body. We may admire it as an instructive approach toward the formation of a many-celled body, but we must look elsewhere for the true upward advance.

The next stage is best illustrated in certain spherical colonies of cells like the tiny green Volvox (now generally regarded as vegetal) of our ponds, or Magosphoera. Here the constituent cells merge their individuality in the common action. We have the first definite many-celled body. It is the type to which a moving close colony of one-celled microbes would soon come. The round surface is well adapted for rolling or spinning along in the water, and, as each little cell earns its own living, it must be at the surface, in contact with the water. Thus a hollow, or fluid-filled, little sphere, like the Volvox, is the natural connecting-link between the microbe and the many-celled body, and may be taken to represent the first important stage in its development.

The next important stage is also very clearly exhibited in nature, and is more or less clearly reproduced in the embryonic development of all animals. We may imagine that the age of microbes was succeeded by an age of these many-celled larger bodies, and the struggle for life entered upon a new phase. The great principle we have already recognised came into play once more. Large numbers of the many-celled bodies shrank from the field of battle, and adopted the method of the plant. They rooted themselves to the floor of the ocean, and developed long arms or lashes for creating a whirlpool movement in the water, and thus bringing the food into their open mouths. Forfeiting mobility, they have, like the plant, forfeited the greater possibilities of progress, and they remain flowering to-day on the floors of our waters, recalling the next phase in the evolution of early life. Such are the hydra, the polyp, the coral, and the sea-anemone. It is not singular that earlier observers could not detect that they were animals, and they were long known in science as "animal-plants" (Zoophytes).

When we look to the common structure of these animals, to find the ancestral type, we must ignore the nerve and muscle-cells which they have developed in some degree. Fundamentally, their body consists of a pouch, with an open mouth, the sides of the pouch consisting of a double layer of cells. In this we have a clue to the next stage of animal development. Take a

94

soft india-rubber ball to represent the first many-celled animal. Press in one half of the ball close upon the other, narrow the mouth, and you have something like the body-structure of the coral and hydra. As this is the course of embryonic development, and as it is so well retained in the lowest groups of the many-celled animals, we take it to be the next stage. The reason for it will become clear on reflection. Division of labour naturally takes place in a colony, and in that way certain cells in the primitive body were confined to the work of digestion. It would be an obvious advantage for these to retire into the interior, leaving the whole external surface free for the adjustment of the animal's relations to the outer world.

Again we must refrain from following in detail the development of this new world of life which branches off in the Archaean ocean. The evolution of the Corals alone would be a lengthy and interesting story. But a word must be said about the jelly-fish, partly because the inexpert will be puzzled at the inclusion of so active an animal, and partly because its story admirably illustrates the principle we are studying. The Medusa really descends from one of the plant-like animals of the early Archaean period, but it has abandoned the ancestral stalk, turned upside down, and developed muscular swimming organs. Its past is betrayed in its embryonic development. As a rule the germ develops into a stalked polyp, out of which the free-swimming Medusa is formed. This return to active and free life must have occurred early, as we find casts of large Medusae in the Cambrian beds. In complete harmony with the principle we laid down, the jelly-fish has gained in nerve and sensitiveness in proportion to its return to an active career.

But this principle is best illustrated in the other branch of the early many-celled animals, which continued to move about in search of food. Here, as will be expected, we have the main stem of the animal world, and, although the successive stages of development are obscure, certain broad lines that it followed are clear and interesting.

It is evident that in a swarming population of such animals the most valuable qualities will be speed and perception. The sluggish Coral needs only sensitiveness enough, and mobility enough, to shrink behind its protecting scales at the approach of danger. In the open water the most speedy and most sensitive will be apt to escape destruction, and have the larger share in breeding the next generation. Imagine a selection on this principle going on for millions of years, and the general result can be conjectured. A very interesting analogy is found in the evolution of the boat. From the clumsy hollowed tree of Neolithic man natural selection, or the need of increasing speed, has developed the elongated, evenly balanced modern boat, with its distinct stem and stern. So in the Archaean ocean the struggle to overtake food, or escape feeders, evolved an elongated two-sided body, with head and tail, and with the oars (cilia) of the one-celled ancestor spread thickly along its flanks. In other words, a body akin to that of the lower water-worms would be the natural result; and this is, in point of fact, the next stage we find in the hierarchy of living nature.

Probably myriads of different types of this worm-like organisation were developed, but such animals leave no trace in the rocks, and we can only follow the development by broad analogies. The lowest flat-worms of to-day may represent some of these early types, and as we ascend the scale of what is loosely called "worm" organisation, we get some instructive suggestions of the way in which the various organs develop. Division of labour continues among the colony of cells which make up the body, and we get distinct nerve-cells, muscle-cells, and digestive cells. The nerve-cells are most useful at the head of an organism which moves through the water, just as the look-out peers from the head of the ship, and there they develop most thickly. By a fresh division of labour some of these cells become especially sensitive to light, some to the chemical qualities of matter, some to movements of the water; we have the beginning of the eyes, the nose, and the ears, as simple little depressions in the skin of the head, lined with these sensitive cells. A muscular gullet arises to protect the digestive tube; a simple drainage channel for waste matter forms under the skin; other channels permit the passage of the fluid food, become (in the higher worms) muscular blood-vessels, and begin to contract—somewhat erratically at first—and drive the blood through the system.

Here, perhaps, are millions of years of development compressed into a paragraph. But the purpose of this work is chiefly to describe the material record of the advance of life in the earth's strata, and show how it is related to great geological changes. We must therefore abstain from endeavouring to trace the genealogy of the innumerable types of animals which were, until recently, collected in zoology under the heading "Worms." It is more pertinent to inquire how the higher classes of animals, which we found in the Cambrian seas, can have arisen from this primitive worm-like population.

The struggle for life in the Archaean ocean would become keener and more exacting with the appearance of each new and more effective type. That is a familiar principle in our industrial world to-day, and we shall find it illustrated throughout our story. We therefore find the various processes of evolution, which we have already seen, now actively at work among the swarming Archaean population, and producing several very distinct types. In some of these struggling organisms speed is developed, together with offensive and defensive weapons, and a line slowly ascends toward the fish, which we will consider later. In others defensive armour is chiefly developed, and we get the lines of the heavy sluggish shell-fish, the Molluscs and Brachiopods, and, by a later compromise between speed and armour, the more active tough-coated Arthropods. In others the plant-principle reappears; the worm-like creature retires from the free-moving life, attaches itself to a fixed base, and becomes the Bryozoan or the Echinoderm. To trace the development of these types in any detail is impossible. The early remains are not preserved. But some clues are found in nature or in embryonic development, and, when the types do begin to be preserved in the rocks, we find the process of evolution plainly at work in them. We will therefore say a few words about the general evolution of each type, and then return to the geological record in the Cambrian rocks.

The starfish, the most familiar representative of the Echinoderms, seems very far removed from the kind of worm-like ancestor we have been imagining, but, fortunately, the very interesting story of the starfish is easily learned from the geological chronicle. Reflect on the flower-like expansion of its arms, and then imagine it mounted on a stalk, mouth side upward, with those arms—more tapering than they now are—waving round the mouth. That, apparently, was the past of the starfish and its cousins. We shall see that the earliest Echinoderms we know are cup-shaped structures on stalks, with a stiff, limy frame and (as in all sessile animals) a number of waving arms round the mouth. In the next geological age the stalk will become a long and flexible arrangement of muscles and plates of chalk, the cup will be more perfectly compacted of chalky plates, and the five arms will taper and branch until they have an almost feathery appearance; and the animal will be considered a "sea-lily" by the early geologist.

The evidence suggests that both the free-moving and the stalked Echinoderms descend from a common stalked Archaean ancestor. Some primitive animal abandoned the worm-like habit, and attached itself, like a polyp, to the floor. Like all such sessile animals, it developed a wreath of arms round the open mouth. The "sea-cucumber" (Holothurian) seems to be a type that left the stalk, retaining the little wreath of arms, before the body was heavily protected and deformed. In the others a strong limy skeleton was developed, and the nerves and other organs were modified in adaptation to the bud-like or flower-like structure. Another branch of the family then abandoned the stalk, and, spreading its arms flat, and gradually developing in them numbers of little "feet" (water-tubes), became the starfish. In the living Comatula we find a star passing through the stalked stage in its early development, when it looks like a tiny sea-lily. The sea-urchin has evolved from the star by folding the arms into a ball. [*]

    * See the section on Echinoderms, by Professor MacBride, in
      the "Cambridge Natural History," I.

The Bryozoa (sea-mats, etc.) are another and lower branch of the primitive active organisms which have adopted a sessile life. In the shell-fish, on the other hand, the principle of armour-plating has its greatest development. It is assuredly a long and obscure way that leads from the ancestral type of animal we have been describing to the headless and shapeless mussel or oyster. Such a degeneration is, however, precisely what we should expect to find in the circumstances. Indeed, the larva, of many of the headless Molluscs have a mouth and eyes, and there is a very common type of larva—the trochosphere—in the Mollusc world which approaches the earlier form of some of the higher worms. The Molluscs, as we shall see, provide some admirable illustrations of the process of evolution. In some of the later fossilised specimens (Planorbis, Paludina, etc.) we can trace the animal as it gradually passes from one species to another. The freshening of the Caspian Sea, which was an outlying part of the Mediterranean quite late in the geological record, seems to have evolved several new genera of Molluscs.

Although, therefore, the remains are not preserved of those primitive Molluscs in which we might see the protecting shell gradually thickening, and deforming the worm-like body, we are not without indications of the process. Two unequal branches of the early wormlike organisms shrank into strong protective shells. The lower branch became the Brachiopods; the more advanced branch the Molluscs. In the Mollusc world, in turn, there are several early types developed. In the Pelecypods (or Lamellibranchs—the mussel, oyster, etc.) the animal retires wholly within its fortress, and degenerates. The Gastropods (snails, etc.) compromise, and retain a certain amount of freedom, so that they degenerate less. The highest group, the Cephalopods, "keep their heads," in the literal sense, and we shall find them advancing from form to form until, in the octopus of a later age, they discard the ancestral shell, and become the aristocrats of the Mollusc kingdom.

The last and most important line that led upward from the chaos of Archaean worms is that of the Arthropods. Its early characteristic was the acquisition of a chitinous coat over the body. Embryonic indications show that this was at first a continuous shield, but a type arose in which the coat broke into sections covering each segment of the body, giving greater freedom of movement. The shield, in fact, became a fine coat of mail. The Trilobite is an early and imperfect experiment of the class, and the larva of the modern king-crab bears witness that it has not perished without leaving descendants. How later Crustacea increase the toughness of the coat by deposits of lime, and lead on to the crab and lobster, and how one early branch invades the land, develops air-breathing apparatus, and culminates in the spiders and insects, will be considered later. We shall see that there is most remarkable evidence connecting the highest of the Arthropods, the insect, with a remote Annelid ancestor.

We are thus not entirely without clues to the origin of the more advanced animals we find when the fuller geological record begins. Further embryological study, and possibly the discovery of surviving primitive forms, of which Central Africa may yet yield a number, may enlarge our knowledge, but it is likely to remain very imperfect. The fossil records of the long ages during which the Mollusc, the Crustacean, and the Echinoderm slowly assumed their characteristic forms are hopelessly lost. But we are now prepared to return to the record which survives, and we shall find the remaining story of the earth a very ample and interesting chronicle of evolution.

CHAPTER VII. THE PASSAGE TO THE LAND

Slender as our knowledge is of the earlier evolution of the Invertebrate animals, we return to our Cambrian population with greater interest. The uncouth Trilobite and its livelier cousins, the sluggish, skulking Brachiopod and Mollusc, the squirming Annelids, and the plant-like Cystids, Corals, and Sponges are the outcome of millions of years of struggle. Just as men, when their culture and their warfare advanced, clothed themselves with armour, and the most completely mailed survived the battle, so, generation after generation, the thicker and harder-skinned animals survived in the Archaean battlefield, and the Cambrian age opened upon the various fashions of armour that we there described. But, although half the story of life is over, organisation is still imperfect and sluggish. We have now to see how it advances to higher levels, and how the drama is transferred from the ocean to a new and more stimulating environment.

The Cambrian age begins with a vigorous move on the part of the land. The seas roll back from the shores of the "lost Atlantis," and vast regions are laid bare to the sun and the rains. In the bays and hollows of the distant shores the animal survivors of the great upheaval adapt themselves to their fresh homes and continue the struggle. But the rivers and the waves are at work once more upon the land, and, as the Cambrian age proceeds, the fringes of the continents are sheared, and the shore-life steadily advances upon the low-lying land. By the end of the Cambrian age a very large proportion of the land is covered with a shallow sea, in which the debris of its surface is deposited. The levelling continues through the next (Ordovician) period. Before its close nearly the whole of the United States and the greater part of Canada are under water, and the new land that had appeared on the site of Europe is also for the most part submerged. The present British Isles are almost reduced to a strip of north-eastern Ireland, the northern extremity of Scotland, and large islands in the south-west and centre of England.

We have already seen that these victories of the sea are just as stimulating, in a different way, to animals as the victories of the land. American geologists are tracing, in a very instructive way, the effect on that early population of the encroachment of the sea. In each arm of the sea is a distinctive fauna. Life is still very parochial; the great cosmopolitans, the fishes, have not yet arrived. As the land is revelled, the arms of the sea approach each other, and at last mingle their waters and their populations, with stimulating effect. Provincial characters are modified, and cosmopolitan characters increase in the great central sea of America. The vast shallow waters provide a greatly enlarged theatre for the life of the time, and it flourishes enormously. Then, at the end of the Ordovician, the land begins to rise once more. Whether it was due to a fresh shrinking of the crust, or to the simple process we have described, or both, we need not attempt to determine; but both in Europe and America there is a great emergence of land. The shore-tracts and the shallow water are narrowed, the struggle is intensified in them, and we pass into the Silurian age with a greatly reduced number but more advanced variety of animals. In the Silurian age the sea advances once more, and the shore-waters expand. There is another great "expansive evolution" of life. But the Silurian age closes with a fresh and very extensive emergence of the land, and this time it will have the most important consequences. For two new things have meantime appeared on the earth. The fish has evolved in the waters, and the plant, at least, has found a footing on the land.

These geological changes which we have summarised and which have been too little noticed until recently in evolutionary studies, occupied 7,000,000 years, on the lowest estimate, and probably twice that period. The impatient critic of evolutionary hypotheses is apt to forget the length of these early periods. We shall see that in the last two or three million years of the earth's story most extraordinary progress has been made in plant and animal development, and can be very fairly traced. How much advance should we allow for these seven or fourteen million years of swarming life and changing environments?

We cannot nearly cover the whole ground of paleontology for the period, and must be content to notice some of the more interesting advances, and then deal more fully with the evolution of the fish, the forerunner of the great land animals.

The Trilobite was the most arresting figure in the Cambrian sea, and its fortunes deserve a paragraph. It reaches its climax in the Ordovician sea, and then begins to decline, as more powerful animals come upon the scene. At first (apparently) an eyeless organism, it gradually develops compound eyes, and in some species the experts have calculated that there were 15,000 facets to each eye. As time goes on, also, the eye stands out from the head on a kind of stalk, giving a wider range of vision. Some of the more sluggish species seem to have been able to roll themselves up, like hedgehogs, in their shells, when an enemy approached. But another branch of the same group (Crustacea) has meantime advanced, and it gradually supersedes the dwindling Trilobites. Toward the close of the Silurian great scorpion-like Crustaceans (Pterygotus, Eurypterus, etc.) make their appearance. Their development is obscure, but it must be remembered that the rocks only give the record of shore-life, and only a part of that is as yet opened by geology. Some experts think that they were developed in inland waters. Reaching sometimes a length of five or six feet, with two large compound eyes and some smaller eye-spots (ocelli), they must have been the giants of the Silurian ocean until the great sharks and other fishes appeared.

The quaint stalked Echinoderm which also we noticed in the Cambrian shallows has now evolved into a more handsome creature, the sea-lily. The cup-shaped body is now composed of a large number of limy plates, clothed with flesh; the arms are long, tapering, symmetrical, and richly fringed; the stalk advances higher and higher, until the flower-like animal sometimes waves its feathery arms from the top of a flexible pedestal composed of millions of tiny chalk disks. Small forests of these sea-lilies adorn the floor of the Silurian ocean, and their broken and dead frames form whole beds of limestone. The primitive Cystids dwindle and die out in the presence of such powerful competitors. Of 250 species only a dozen linger in the Silurian strata, though a new and more advanced type—the Blastoid—holds the field for a time. It is the age of the Crinoids or sea-lilies. The starfish, which has abandoned the stalk, does not seem to prosper as yet, and the brittle-star appears. Their age will come later. No sea-urchins or sea-cucumbers (which would hardly be preserved) are found as yet. It is precisely the order of appearance which our theory of their evolution demands.

The Brachiopods have passed into entirely new and more advanced species in the many advances and retreats of the shores, but the Molluscs show more interesting progress. The commanding group from the start is that of the Molluscs which have "kept their head," the Cephalopods, and their large shells show a most instructive evolution. The first great representative of the tribe is a straight-shelled Cephalopod, which becomes "the tyrant and scavenger of the Silurian ocean" (Chamberlin). Its tapering, conical shell sometimes runs to a length of fifteen feet, and a diameter of one foot. It would of itself be an important evolutionary factor in the primitive seas, and might explain more than one advance in protective armour or retreat into heavy shells. As the period advances the shell begins to curve, and at last it forms a close spiral coil. This would be so great an advantage that we are not surprised to find the coiled type (Goniatites) gain upon and gradually replace the straight-shelled types (Orthoceratites). The Silurian ocean swarms with these great shelled Cephalopods, of which the little Nautilus is now the only survivor.

We will not enlarge on the Sponges and Corals, which are slowly advancing toward the higher modern types. Two new and very powerful organisms have appeared, and merit the closest attention. One is the fish, the remote ancestor of the birds and mammals that will one day rule the earth. The other may be the ancestor of the fish itself, or it may be one of the many abortive outcomes and unsuccessful experiments of the stirring life of the time. And while these new types are themselves a result of the great and stimulating changes which we have reviewed and the incessant struggle for food and safety, they in turn enormously quicken the pace of development. The Dreadnought appears in the primitive seas; the effect on the fleets of the world of the evolution of our latest type of battleship gives us a faint idea of the effect, on all the moving population, of the coming of these monsters of the deep. The age had not lacked incentives to progress; it now obtains a more terrible and far-reaching stimulus.

To understand the situation let us see how the battle of land and sea had proceeded. The Devonian Period had opened with a fresh emergence of the land, especially in Europe, and great inland seas or lakes were left in the hollows. The tincture of iron which gives a red colour to our characteristic Devonian rocks, the Old Red Sandstone, shows us that the sand was

108

deposited in inland waters. The fish had already been developed, and the Devonian rocks show it swarming, in great numbers and variety, in the enclosed seas and round the fringe of the continents.

The first generation was a group of strange creatures, half fish and half Crustacean, which are known as the Ostracoderms. They had large armour-plated heads, which recall the Trilobite, and suggest that they too burrowed in the mud of the sea or (as many think) of the inland lakes, making havoc among the shell-fish, worms, and small Crustacea. The hind-part of their bodies was remarkably fish-like in structure. But they had no backbone—though we cannot say whether they may not have had a rod of cartilage along the back—and no articulated jaws like the fish. Some regard them as a connecting link between the Crustacea and the fishes, but the general feeling is that they were an abortive development in the direction of the fish. The sharks and other large fishes, which have appeared in the Silurian, easily displace these clumsy and poor-mouthed competitors One almost thinks of the aeroplane superseding the navigable balloon.

Of the fishes the Arthrodirans dominated the inland seas (apparently), while the sharks commanded the ocean. One of the Arthrodirans, the Dinichthys ("terrible fish"), is the most formidable fish known to science. It measured twenty feet from snout to tail. Its monstrous head, three feet in width, was heavily armoured, and, instead of teeth, its great jaws, two feet in length, were sharpened, and closed over the victim like a gigantic pair of clippers. The strongly plated heads of these fishes were commonly a foot or two feet in width. Life in the waters became more exacting than ever. But the Arthrodirans were unwieldy and sluggish, and had to give way before more progressive types. The toothed shark gradually became the lord of the waters.

The early shark ate, amongst other things, quantities of Molluscs and Brachiopods. Possibly he began with Crustacea; in any case the practice of crunching shellfish led to a stronger and stronger development of the hard plate which lined his mouth. The prickles of the plate grew larger and harder, until—as may be seen to-day in the mouth of a young shark—the cavity was lined with teeth. In the bulk of the Devonian sharks these developed into what are significantly called "pavement teeth." They were solid plates of enamel, an inch or an inch and a half in width, with which the monster ground its enormous meals of Molluscs, Crustacea, sea-weed, etc. A new and stimulating element had come into the life of the invertebrate world. Other sharks snapped larger victims, and developed the teeth on the edges of their jaws, to the sacrifice of the others, until we find these teeth in the course of time solid triangular masses of enamel, four or five inches long, with saw-like edges. Imagine these terrible mouths—the shears of the Arthrodiran, and the grindstones and terrible crescents of the giant sharks— moving speedily amongst the crowded inhabitants of the waters, and it is easy to see what a stimulus to the attainment of speed and of protective devices was given to the whole world of the time.

What was the origin of the fish? Here we are in much the same position as we were in regard to the origin of the higher Invertebrates. Once the fish plainly appears upon the scene it is found to be undergoing a process of evolution like all other animals. The vast majority of our fishes have bony frames (or are Teleosts); the fishes of the Devonian age nearly all have frames of cartilage, and we know from embryonic development that cartilage is the first stage in the formation of bone. In the teeth and tails, also, we find a gradual evolution toward the higher types. But the earlier record is, for reasons I have already given, obscure; and as my purpose is rather to discover the agencies of evolution than to strain slender evidence in drawing up pedigrees, I need only make brief reference to the state of the problem.

Until comparatively recent times the animal world fell into two clearly distinct halves, the Vertebrates and the Invertebrates. There were several anatomical differences between the two provinces, but the most conspicuous and most puzzling was the backbone. Nowhere in living nature or in the rocks was any intermediate type known between the backboned and the non-backboned animal. In the course of the nineteenth century, however, several animals of an intermediate type were found. The sea-squirt has in its early youth the line of cartilage through the body which, in embryonic development, represents the first stage of the backbone; the lancelet and the Appendicularia have a rod of cartilage throughout life; the "acorn-headed worm" shows traces of it. These are regarded as surviving specimens of various groups of animals which, in early times, fell between the

Invertebrate and Vertebrate worlds, and illustrate the transition.

With their aid a genealogical tree was constructed for the fish. It was assumed that some Cambrian or Silurian Annelid obtained this stiffening rod of cartilage. The next advantage—we have seen it in many cases—was to combine flexibility with support. The rod was divided into connected sections (vertebrae), and hardened into bone. Besides stiffening the body, it provided a valuable shelter for the spinal cord, and its upper part expanded into a box to enclose the brain. The fins were formed of folds of skin which were thrown off at the sides and on the back, as the animal wriggled through the water. They were of use in swimming, and sections of them were stiffened with rods of cartilage, and became the pairs of fins. Gill slits (as in some of the highest worms) appeared in the throat, the mouth was improved by the formation of jaws, and—the worm culminated in the shark.

Some experts think, however, that the fish developed directly from a Crustacean, and hold that the Ostracoderms are the connecting link. A close discussion of the anatomical details would be out of place here, [*] and the question remains open for the present. Directly or indirectly, the fish is a descendant of some Archaean Annelid. It is most probable that the shark was the first true fish-type. There are unrecognisable fragments of fishes in the Ordovician and Silurian rocks, but the first complete skeletons (Lanarkia, etc.) are of small shark- like creatures, and the low organisation of the group to which the shark belongs, the Elasmobranchs, makes it probable that they are the most primitive. Other remains (Palaeospondylus) show that the fish-like lampreys had already developed.

Two groups were developed from the primitive fish, which have great interest for us. Our next step, in fact, is to trace the passage of the fish from the water to the land, one of the most momentous chapters in the story of life. To that incident or accident of primitive life we owe our own existence and the whole development of the higher types of animals. The advance of natural history in modern times has made this passage to the land easy to understand. Not only does every frog reenact it in the course of its development, but we know many fishes that can live out of water. There is an Indian perch—called the "climbing perch," but it has only once been seen by a European to climb a tree—which crosses the fields in search of another pool, when its own pool is evaporating. An Indian marine fish (Periophthalmus) remains hunting on the shore when the tide goes out. More important still, several fishes have lungs as well as gills. The Ceratodus of certain Queensland rivers has one lung; though, I was told by the experts in Queensland, it is not a "mud-fish," and never lives in dry mud. However, the Protopterus of Africa and the Lepidosiren of South America have two lungs, as well as gills, and can live either in water or, in the dry season, on land.

When the skeletons of fishes of the Ceratodus type were discovered in the Devonian rocks, it was felt that we had found the fish-ancestor of the land Vertebrates, but a closer anatomical examination has made this doubtful. The Devonian lung-fish has characters which do not seem to lead on to the Amphibia. The same general cause probably led many groups to leave the water, or adapt themselves to living on land as well as in water, and the abundant Dipoi or Dipneusts ("double-breathers") of the Devonian lakes are one of the chief of these groups, which have luckily left descendants to our time. The ancestors of the Amphibia are generally sought amongst the Crossopterygii, a very large group of fishes in Devonian times, with very few representatives to-day.

It is more profitable to investigate the process itself than to make a precarious search for the actual fish, and, fortunately, this inquiry is more hopeful. The remains that we find make it probable that the fish left the water about the beginning of the Devonian or the end of the Silurian. Now this period coincides with two circumstances which throw a complete light on the step; one is the great rise of the land, catching myriads of fishes in enclosed inland seas, and the other is the appearance of formidable carnivores in the waters. As the seas evaporated [*] and the great carnage proceeded, the land, which was already covered with plants and inhabited by insects, offered a safe retreat for such as could adopt it. Emigration to the land had been going on for ages, as we shall see. Curious as it must seem to the inexpert, the fishes, or some of them, were better prepared than most other animals to leave the water. The chief requirement was a lung, or interior bag, by which the air could be brought into close contact with the absorbing blood vessels. Such a bag, broadly speaking, most of the fishes possess in their floating-bladder: a bag of gas, by compressing or expanding which they alter their specific gravity in the water. In some fishes it is double; in some it is supplied with blood-vessels; in some it is connected by a tube with the gullet, and therefore with the atmosphere.

```
    * It is now usually thought that the inland seas
were the
    theatre of the passage to land. I must point out,
however,
    that the wide distribution of our Dipneusts, in
Australia,
    tropical Africa, and South America, suggests that
they were
    marine though they now live in fresh water. But we
shall see
    that a continent united the three regions at one
time, and
    it may afford some explanation.
```

Thus we get very clear suggestions of the transition from water to land. We must, of course, conceive it as a slow and gradual adaptation. At first there may have been a rough contrivance for deriving oxygen directly and partially from the atmosphere, as the water of the lake became impure. So important an advantage would be fostered, and, as the inland sea became smaller, or its population larger or fiercer, the fishes with a sufficiently developed air-breathing apparatus passed to the land, where, as yet, they would find no serious enemy. The fact is beyond dispute; the theory of how it occurred is plausible enough; the consequences were momentous. Great changes were preparing on the land, and in a comparatively short time we shall find its new inhabitant subjected to a fierce test of circumstances that will carry it to an enormously higher level than life had yet reached.

I have said that the fact of this transition to the land is beyond dispute. The evidence is very varied, but need not all be enlarged upon here. The widespread Dipneust fishes of the Devonian rocks bear strong witness to it, and the appearance of the Amphibian immediately afterwards makes it certain. The development of the frog is a reminiscence of it, on the lines of the embryonic law which we saw earlier. An animal, in its individual development, more or less reproduces the past phases of its ancestry. So the free-swimming jelly-fish begins life as a fixed polyp; a kind of star-fish (Comatula) opens its career as a stalked sea-lily; the gorgeous dragon-fly is at first an uncouth aquatic animal, and the ethereal butterfly a worm-like creature. But the most singular and instructive of all these embryonic reminiscences of the past is found in the fact that all the higher land-animals of to-day clearly reproduce a fish-stage in their embryonic development.

In the third and fourth weeks of development the human embryo shows four (closed) slits under the head, with corresponding arches. The bird, the dog, the horse—all the higher land animals, in a word, pass through the same phase. The suggestion has been made that these structures do not recall the gill-slits and gill-arches of the fish, but are folds due to the packing of the embryo in the womb. In point of fact, they appear just at the time when the human embryo is only a fifth of an inch long, and there is no such compression. But all doubt as to their interpretation is dispelled when we remove the skin and examine the heart and blood-vessels. The heart is up in the throat, as in the fish, and has only two chambers, as in the fish (not four, as in the bird and mammal); and the arteries rise in five pairs of arches over the swellings in the throat, as they do in the lower fish, but do not in the bird and mammal. The arrangement is purely temporary—lasting only a couple of weeks in the human embryo—and purposeless. Half these arteries will disappear again. They quite plainly exist to supply fine blood-vessels for breathing at the gill-clefts, and are never used, for the embryo does not breathe, except through the mother. They are a most instructive reminder of the Devonian fish which quitted its element and became the ancestor of all the birds and mammals of a later age.

Several other features of man's embryonic development—the budding of the hind limbs high up, instead of at the base of, the vertebral column, the development of the ears, the nose, the jaws, etc.—have the same lesson, but the one detailed illustration will suffice. The millions of years of stimulating change and struggle which we have summarised have resulted in the production of a fish which walks on four limbs (as the South American mud-fish does to-day), and breathes the atmosphere.

We have been quite unable to follow the vast changes which have meantime taken place in its organisation. The eyes, which were mere pits in the skin, lined with pigment cells, in the early worm, now have a crystalline lens to concentrate the light and define objects on the nerve. The ears, which were at first similar sensitive pits in the skin, on which lay a little stone whose movements gave the animal some sense of direction, are now closed vesicles in the skull, and begin to be sensitive to waves of sound. The nose, which was at first two blind, sensitive pits in the skin of the head, now consists of two nostrils opening into the mouth, with an olfactory nerve spreading richly over the passages. The brain, which was a mere clump of nerve-cells connecting the rough sense-impressions, is now a large and intricate structure, and already exhibits a little of that important region (the cerebrum) in which the varied images of the outside world are combined. The heart, which was formerly was a mere swelling of a part of one of the blood-vessels, now has two chambers.

We cannot pursue these detailed improvements of the mechanism, as we might, through the ascending types of animals. Enough if we see more or less clearly how the changes in the face of the earth and the rise of its successive dynasties of carnivores have stimulated living things to higher and higher levels in the primitive ocean. We pass to the clearer and far more important story of life on land, pursuing the fish through its continuous adaptations to new conditions until, throwing out side-branches as it progresses, it reaches the height of bird and mammal life.

CHAPTER VIII. THE COAL-FOREST

With the beginning of life on land we open a new and more important volume of the story of life, and we may take the opportunity to make clearer certain principles or processes of development which we may seem hitherto to have taken for granted. The evolutionary work is too often a mere superficial description of the strange and advancing classes of plants and animals which cross the stage of geology. Why they change and advance is not explained. I have endeavoured to supply this explanation by putting the successive populations of the earth in their respective environments, and showing the continuous and stimulating effect on them of changes in those environments. We have thus learned to decipher some lines of the decalogue of living nature. "Thou shalt have a thick armour," "Thou shalt be speedy," "Thou shalt shelter from the more powerful," are some of the laws of primeval life. The appearance of each higher and more destructive type enforces them with more severity; and in their observance animals branch outward and upward into myriads of temporary or permanent forms.

But there is no consciousness of law and no idea of evading danger. There is not even some mysterious instinct "telling" the animal, as it used to be said, to do certain things. It is, in fact, not strictly accurate to say that a certain change in the environment stimulates animals to advance. Generally speaking, it does not act on the advancing at all, but on the non-advancing, which it exterminates. The procedure is simple, tangible, and unconscious. Two invading arms of the sea meet and pour together their different waters and populations. The habits, the foods, and the enemies of many types of animals are changed; the less fit for the new environment die first, the more fit survive longest and breed most of the new generation. It is so with men when they migrate to a more exacting environment, whether a dangerous trade or a foreign clime. Again, take the case of the introduction of a giant Cephalopod or fish amongst a population of Molluscs and Crustacea. The toughest, the speediest, the most alert, the most retiring, or the least conspicuous, will be the most apt to survive and breed. In hundreds or thousands of generations there will be an enormous improvement in the armour, the speed, the sensitiveness, the hiding practices, and the protective colours, of the animals which are devoured. The "natural selection of the fittest" really means the "natural destruction of the less fit."

The only point assumed in this is that the young of an animal or plant tend to differ from each other and from their parents. Darwin was content to take this as a fact of common observation, as it obviously is, but later science has thrown some light on the causes of these variations. In the first place, the germs in the parent's body may themselves be subject to struggle and natural selection, and not share equally in the food-supply. Then, in the case of the higher animals (or the majority of animals), there is a clear source of variation in the fact that the mature germ is formed of certain elements from two different parents, four grandparents, and so on. In the case of the lower animals the germs and larvae float independently in the water, and are exposed to many influences. Modern embryologists have found, by experiment, that an alteration of the temperature or the chemical considerable effect on eggs and larvae. Some recent experiments have shown that such changes may even affect the eggs in the mother's ovary. These discoveries are very important and suggestive, because the geological changes which we are studying are especially apt to bring about changes of temperature and changes in the freshness or saltiness of water.

Evolution is, therefore, not a "mere description" of the procession of living things; it is to a great extent an explanation of the procession. When, however, we come to apply these general principles to certain aspects of the advance in organisation we find fundamental differences of opinion among biologists, which must be noted. As Sir E. Ray Lankester recently said, it is not at all true that Darwinism is questioned in zoology to-day. It is true only that Darwin was not omniscient or infallible, and some of his opinions are disputed.

Let me introduce the subject with a particular instance of evolution, the flat-fish. This animal has been fitted to survive the terrible struggle in the seas by acquiring such a form that it can lie almost unseen upon the floor of the ocean. The eye on the under side of the body would thus be useless, but a glance at a sole or plaice in a fishmonger's shop will show that this eye has worked upward to the top of the head. Was the eye shifted by the effort and straining of the fish, inherited and increased slightly in each generation? Is the explanation rather that those fishes in each generation survived and bred which happened from birth to have a slight variation in that direction, though they did not inherit the effect of the parent's effort to strain the eye? Or ought we to regard this change of structure as brought about by a few abrupt and considerable variations on the part of the young? There you have the three great schools which divide modern evolutionists: Lamarckism, Weismannism, and Mendelism (or Mutationism). All are Darwinians. No one doubts that the flat-fish was evolved from an ordinary fish—the flat-fish is an ordinary fish in its youth—or that natural selection (enemies) killed off the old and transitional types and overlooked (and so favoured) the new. It will be seen that the language used in this volume is not the particular language of any one of these schools. This is partly because I wish to leave seriously controverted questions open, and partly from a feeling of compromise, which I may explain. [*]

* Of recent years another compromise has been proposed
between the Lamarckians and Weismannists. It would say that
the efforts of the parent and their effect on the position
of the eye—in our case—are not inherited, but might be of
use in sheltering an embryonic variation in the direction of
a displaced eye.

First, the plain issue between the Mendelians and the other two schools—whether the passage from species to species is brought about by a series of small

122

variations during a long period or by a few large variations (or "mutations") in a short period—is open to an obvious compromise. It is quite possible that both views are correct, in different cases, and quite impossible to find the proportion of each class of cases. We shall see later that in certain instances where the conditions of preservation were good we can sometimes trace a perfectly gradual advance from species to species. Several shellfish have been traced in this way, and a sea-urchin in the chalk has been followed, quite gradually, from one end of a genus to the other. It is significant that the advance of research is multiplying these cases. There is no reason why we may not assume most of the changes of species we have yet seen to have occurred in this way. In fact, in some of the lower branches of the animal world (Radiolaria, Sponges, etc.) there is often no sharp division of species at all, but a gradual series of living varieties.

On the other hand we know many instances of very considerable sudden changes. The cases quoted by Mendelists generally belong to the plant world, but instances are not unknown in the animal world. A shrimp (Artemia) was made to undergo considerable modification, by altering the proportion of salt in the water in which it was kept. Butterflies have been made to produce young quite different from their normal young by subjecting them to abnormal temperature, electric currents, and so on; and, as I said, the most remarkable effects have been produced on eggs and embryos by altering the chemical and physical conditions. Rats—I was informed by the engineer in charge of the refrigerating room on an Australian liner—very quickly became adapted to the freezing temperature by developing long hair. All that we have seen of the past changes in the environment of animals makes it probable that these larger variations often occur. I would conclude, therefore, that evolution has proceeded continuously (though by no means universally) through the ages, but there were at times periods of more acute change with correspondingly larger changes in the animal and plant worlds.

In regard to the issue between the Lamarckians and Weismannists—whether changes acquired by the parent are inherited by the young—recent experiments again suggest something of a compromise. Weismann says that the body of the parent is but the case containing the germ-plasm, so that all modifications of the living parent body perish with it, and do not affect the germ, which builds the next generation. Certainly, when we reflect that the 70,000 ova in the human mother's ovary seem to have been all formed in the first year of her life, it is difficult to see how modifications of her muscles or nerves can affect them. Thus we cannot hope to learn anything, either way, by cutting off the tails of cows, and experiments of that kind. But it is acknowledged that certain diseases in the blood, which nourishes the germs, may affect them, and recent experimenters have found that they can reach and affect the germs in the body by other agencies, and so produce inherited modifications in the parent. [*] If this claim is sustained and enlarged, it may be concluded that the greater changes of environment which we find in the geological chronicle may have had a considerable influence of this kind.

* See a paper read by Professor Bourne to the Zoological
    Section of the British Association, 1910. It must be
    understood that when I speak of Weismannism I do not refer
    to this whole theory of heredity, which, he acknowledges,
    has few supporters. The Lamarckian view is represented in
    Britain by Sir W. Turner and Professor Darwin. In other
    countries it has a larger proportion of distinguished
    supporters. On the whole subject see Professor J. A.
    Thomson's "Heredity" (1909), Dewar and Finn's "Making of
    Species" (1909—a Mendelian work), and, for essays by the
    leaders of each school, "Darwinism and Modern Science"
    (1909).

The general issue, however, must remain open. The Lamarckian and Weismannist theories are rival interpretations of past events, and we shall not find it necessary to press either. When the fish comes to live on land, for instance, it develops a bony limb out of its fin. The Lamarckian says that the throwing of the weight of the body on the main stem of the fin strengthens it, as practice strengthens the boxer's arm, and the effect is inherited and increased in each generation, until at last the useless paddle of the fin dies away and the main stem has become a stout, bony column. Weismann says that the individual modification, by use in walking, is not inherited, but those young are favoured which have at birth a variation in the strength of the stem of the fin. As each of these interpretations is, and must remain, purely theoretical, we will be content to tell the facts in such cases. But these brief remarks will enable the reader to understand in what precise sense the facts we record are open to controversy.

Let us return to the chronicle of the earth. We had reached the Devonian age, when large continents, with great inland seas, existed in North America, north-west Europe, and north Asia, probably connected by a continent across the North Atlantic and the Arctic region. South America and South Africa were emerging, and a continent was preparing to stretch from Brazil, through South Africa and the Antarctic, to Australia and India. The expanse of land was, with many oscillations, gaining on the water, and there was much emigration to it from the over-populated seas. When the fish went on land in the Devonian, it must have found a diet (insects, etc.) there, and the insects must have been preceded by a plant population. We have first, therefore, to consider the evolution of the plant, and see how it increases in form and number until it covers the earth with the luxuriant forests of the Carboniferous period.

The plant world, we saw, starts, like the animal world, with a great kingdom of one-celled microscopic representatives, and the same principles of development, to a great extent, shape it into a large variety of forms. Armour-plating has a widespread influence among them. The graceful Diatom is a morsel of plasm enclosed in a flinty box, often with a very pretty arrangement of the pores and markings. The Desmid has a coat of cellulose, and a less graceful

126

coat of cellulose encloses the Peridinean. Many of these minute plants develop locomotion and a degree of sensitiveness (Diatoms, Peridinea, Euglena, etc.). Some (Bacteria) adopt animal diet, and rise in power of movement and sensitiveness until it is impossible to make any satisfactory distinction between them and animals. Then the social principle enters. First we have loose associations of one-celled plants in a common bed, then closer clusters or many-celled bodies. In some cases (Volvox) the cluster, or the compound plant, is round and moves briskly in the water, closely resembling an animal. In most cases, the cells are connected in chains, and we begin to see the vague outline of the larger plant.

When we had reached this stage in the development of animal life, we found great difficulty in imagining how the chief lines of the higher Invertebrates took their rise from the Archaean chaos of early many-celled forms. We have an even greater difficulty here, as plant remains are not preserved at all until the Devonian period. We can only conclude, from the later facts, that these primitive many-celled plants branched out in several different directions. One section (at a quite unknown date) adopted an organic diet, and became the Fungi; and a later co-operation, or life-partnership, between a Fungus and a one-celled Alga led to the Lichens. Others remained at the Alga-level, and grew in great thickets along the sea bottoms, no doubt rivalling or surpassing the giant sea-weeds, sometimes 400 feet long, off the American coast to-day. Other lines which start from the level of the primitive many-celled Algae develop into the Mosses

(Bryophyta), Ferns (Pteridophyta), Horsetails (Equisetalia), and Club-mosses (Lycopodiales). The mosses, the lowest group, are not preserved in the rocks; from the other three classes will come the great forests of the Carboniferous period.

The early record of plant-life is so poor that it is useless to speculate when the plant first left the water. We have somewhat obscure and disputed traces of ferns in the Ordovician, and, as they and the Horsetails and Club-mosses are well developed in the Devonian, we may assume that some of the sea-weeds had become adapted to life on land, and evolved into the early forms of the ferns, at least in the Cambrian period. From that time they begin to weave a mantle of sombre green over the exposed land, and to play a most important part in the economy of nature.

We saw that at the beginning of the Devonian there was a considerable rise of the land both in America and Europe, but especially in Europe. A distant spectator at that time would have observed the rise of a chain of mountains in Scotland and a general emergence of land north-western Europe. A continent stretched from Ireland to Scandinavia and North Russia, while most of the rest of Europe, except large areas of Russia, France, Germany, and Turkey, was under the sea. Where we now find our Alps and Pyrenees towering up to the snow-line there were then level stretches of ocean. Even the north-western continent was scooped into great inland seas or lagoons, which stretched from Ireland to Scandinavia, and, as we saw, fostered the development of the fishes.

As the Devonian period progressed the sea gained on the land, and must have restricted the growth of vegetation, but as the lake deposits now preserve the remains of the plants which grow down to their shores, or are washed into them, we are enabled to restore the complexion of the landscape. Ferns, generally of a primitive and generalised character, abound, and include the ferns such as we find in warm countries to-day. Horsetails and Club-mosses already grow into forest-trees. There are even seed-bearing ferns, which give promise of the higher plants to come, but as yet nothing approaching our flower and fruit-bearing trees has appeared. There is as yet no certain indication of the presence of Conifers. It is a sombre and monotonous vegetation, unlike any to be found in any climate to-day.

We will look more closely into its nature presently. First let us see how these primitive types of plants come to form the immense forests which are recorded in our coal-beds. Dr. Russel Wallace has lately represented these forests, which have, we shall see, had a most important influence on the development of life, as somewhat mysterious in their origin. If, however, we again consult the geologist as to the changes which were taking place in the distribution of land and water, we find a quite natural explanation. Indeed, there are now distinguished geologists (e.g. Professor Chamberlin) who doubt if the Coal-forests were so exceptionally luxuriant as is generally believed. They think that the vegetation may not have been more dense than in some other ages, but that there may have been exceptionally good conditions for preserving the dead trees. We shall see that there were; but, on the whole, it seems probable that during some hundreds of thousands of years remarkably dense forests covered enormous stretches of the earth's surface, from the Arctic to the Antarctic.

The Devonian period had opened with a rise of the land, but the sea eat steadily into it once more, and, with some inconsiderable oscillations of the land, regained its territory. The latter part of the Devonian and earlier part of the Carboniferous were remarkable for their great expanses of shallow water and low-lying land. Except the recent chain of hills in Scotland we know of no mountains. Professor Chamberlin calculates that 20,000,000, or 30,000,000 square miles of the present continental surface of Europe and America were covered with a shallow sea. In the deeper and clearer of these waters the earliest Carboniferous rocks, of limestone, were deposited. The "millstone grit," which succeeds the "limestone," indicates shallower water, which is being rapidly filled up with the debris of the land. In a word, all the indications suggest the early and middle Carboniferous as an age of vast swamps, of enormous stretches of land just above or below the sea-level, and changing repeatedly from one to the other. Further, the climate was at the time—we will consider the general question of climate later—moist and warm all over the earth, on account of the great proportion of sea-surface and the absence of high land (not to speak of more disputable causes).

These were ideal conditions for the primitive vegetation, and it spread over the swamps with great vigour. To say that the Coal-forests were masses of Ferns, Horsetails, and Club-mosses is a lifeless and misleading expression. The Club-mosses, or Lycopodiales, were massive trees, rising sometimes to a height of 120 feet, and probably averaging about fifty feet in height and one or two feet in diameter. The largest and most abundant of them, the Sigillaria, sent up a scarred and fluted trunk to a height of seventy or a hundred feet, without a branch, and was crowned with a bunch of its long, tapering leaves. The Lepidodendron, its fellow monarch of the forest, branched at the summit, and terminated in clusters of its stiff, needle-like leaves, six' or seven inches long, like enormous exaggerations of the little cones at the ends of our Club-mosses to-day. The Horsetails, which linger in their dwarfed descendants by our streams to-day, and at their exceptional best (in a part of South America) form slender stems about thirty feet high, were then forest-trees, four to six feet in circumference and sometimes ninety feet in height. These Calamites probably rose in dense thickets from the borders of the lakes, their stumpy leaves spreading in whorls at every joint in their hollow stems. Another extinct tree, the Cordaites, rivalled the Horsetails and Club-mosses in height, and its showers of long and extraordinary leaves, six feet long and six inches in width, pointed to the higher plant world that was to come. Between these gaunt towering trunks the graceful tree-ferns spread their canopies at heights of twenty, forty, and even sixty feet from the ground, and at the base was a dense undergrowth of ferns and fern-like seed-plants. Mosses may have carpeted the moist ground, but nothing in the nature of grass or flowers had yet appeared.

Imagine this dense assemblage of dull, flowerless trees pervaded by a hot, dank atmosphere, with no change of seasons, with no movement but the flying of large and primitive insects among the trees and the stirring of the ferns below by some passing giant salamander, with no song of bird and no single streak of white or red or blue drawn across the changeless sombre green, and you have some idea of the character of the forests that are compressed into our seams of coal. Imagine these forests spread from Spitzbergen to Australia and even, according to the south polar expeditions, to the Antarctic, and from the United States to Europe, to Siberia, and to China, and prolonged during some hundreds of thousands of years, and you begin to realise that the Carboniferous period prepared the land for the coming dynasties of animals. Let some vast and terrible devastation fall upon this luxuriant world, entombing the great multitude of its imperfect forms and selecting the higher types for freer life, and the earth will pass into a new age.

But before we describe the animal inhabitants of these forests, the part that the forests play in the story of life, and the great cataclysm which selects the higher types from the myriads of forms which the warm womb of the earth has poured out, we must at least glance at the evolutionary position of the Carboniferous plants themselves. Do they point downward to lower forms, and upward to higher forms, as the theory of evolution requires? A close inquiry into this would lead us deep into the problems of the

132

modern botanist, but we may borrow from him a few of the results of the great labour he has expended on the subject within the last decade.

Just as the animal world is primarily divided into Invertebrates and Vertebrates, the plant world is primarily divided into a lower kingdom of spore-bearing plants (the Cryptogams) and an upper kingdom of seed-bearing plants (the Phanerogams). Again, just as the first half of the earth's story is the age of Invertebrate animals, so it is the age of Cryptogamous plants. So far evolution was always justified in the plant record. But there is a third parallel, of much greater interest. We saw that at one time the evolutionist was puzzled by the clean division of animals into Invertebrate and Vertebrate, and the sudden appearance of the backbone in the chronicle: he was just as much puzzled by the sharp division of our plants into Cryptogams and Phanerogams, and the sudden appearance of the latter on the earth during the Coal-forest period. And the issue has been a fresh and recent triumph for evolution.

Plants are so well preserved in the coal that many years of microscopic study of the remains, and patient putting-together of the crushed and scattered fragments, have shown the Carboniferous plants in quite a new light. Instead of the Coal-forest being a vast assemblage of Cryptogams, upon which the higher type of the Phanerogam is going suddenly to descend from the clouds, it is, to a very great extent, a world of plants that are struggling upward, along many paths, to the higher level. The characters of the Cryptogam and Phanerogam are so mixed up in it that, although the

133

special lines of development are difficult to trace, it is one massive testimony to the evolution of the higher from the lower. The reproductive bodies of the great Lepidodendra are sometimes more like seeds than spores, while both the wood and the leaves of the Sigillaria have features which properly belong to the Phanerogam. In another group (called the Sphenophyllales) the characters of these giant Clubmosses are blended with the characters of the giant Horsetails, and there is ground to think that the three groups have descended from an earlier common ancestor.

Further, it is now believed that a large part of what were believed to be Conifers, suddenly entering from the unknown, are not Conifers at all, but Cordaites. The Cordaites is a very remarkable combination of features that are otherwise scattered among the Cryptogams, Cycads, and Conifers. On the other hand, a very large part of what the geologist had hitherto called "Ferns" have turned out to be seed-bearing plants, half Cycad and half Fern. Numbers of specimens of this interesting group—the Cycadofilices (cycad-ferns) or Pteridosperms (seed-ferns)—have been beautifully restored by our botanists. [*] They have afforded a new and very plausible ancestor for the higher trees which come on the scene toward the close of the Coal-forests, while their fern-like characters dispose botanists to think that they and the Ferns may be traced to a common ancestor. This earlier stage is lost in those primitive ages from which not a single leaf has survived in the rocks. We can only say that it is probable that the Mosses, Ferns, Lycopods, etc., arose independently from the primitive level. But the higher and more important development is now much clearer. The Coal-forest is not simply a kingdom of Cryptogams. It is a world of aspiring and mingled types. Let it be subjected to some searching test, some tremendous spell of adversity, and we shall understand the emergence of the higher types out of the luxuriant profusion and confusion of forms.

* See, especially, D. H. Scott, "Studies of Fossil Botany"
    (2nd ed., 1908), and "The Evolution of Plants" (1910—small
        popular manual).

CHAPTER IX. THE ANIMALS OF THE COAL-FOREST

We have next to see that when this period of searching adversity comes—as it will in the next chapter—the animal world also offers a luxuriant variety of forms from which the higher types may be selected. This, it need hardly be said, is just what we find in the geological record. The fruitful, steaming, rich-laden earth now offered tens of millions of square miles of pasture to vegetal feeders; the waters, on the other hand, teemed with gigantic sharks, huge Cephalopods, large scorpion-like and lobster-like animals, and shoals of armour-plated, hard-toothed fishes. Successive swarms of vegetarians—Worms, Molluscs, etc.—followed the plant on to the land; and swarms of carnivores followed the vegetarians, and assumed strange, new forms in adaptation to land-life. The migration had probably proceeded throughout the Devonian period, especially from the calmer shores of the inland seas. By the middle of the Coal-forest period there was a very large and varied animal population on the land. Like the plants, moreover, these animals were of an intermediate and advancing nature. No bird or butterfly yet flits from tree to tree; no mammal rears its young in the shelter of the ferns. But among the swarming population are many types that show a beginning of higher organisation, and there is a rich and varied material provided for the coming selection.

The monarch of the Carboniferous forest is the Amphibian. In that age of spreading swamps and "dim, watery woodlands," the stupid and sluggish Amphibian finds his golden age, and, except perhaps the scorpion, there is no other land animal competent to dispute his rule. Even the scorpion, moreover, would not find the Carboniferous Amphibian very vulnerable. We must not think of the smooth-skinned frogs and toads and innocent newts which to-day represent the fallen race of the Amphibia. They were then heavily armoured, powerfully armed, and sometimes as large as alligators or young crocodiles. It is a characteristic of advancing life that a new type of organism has its period of triumph, grows to enormous proportions, and spreads into many different types, until the next higher stage of life is reached, and it is dethroned by the new-comers.

The first indication—apart from certain disputed impressions in the Devonian—of the land-vertebrate is the footprint of an Amphibian on an early

Carboniferous mud-flat. Hardened by the sun, and then covered with a fresh deposit when it sank beneath the waters, it remains to-day to witness the arrival of the five-toed quadruped who was to rule the earth. As the period proceeds, remains are found in great abundance, and we see that there must have been a vast and varied population of the Amphibia on the shores of the Carboniferous lagoons and swamps. There were at least twenty genera of them living in what is now the island of Britain, and was then part of the British-Scandinavian continent. Some of them were short and stumpy creatures, a few inches long, with weak limbs and short tails, and broad, crescent-shaped heads, their bodies clothed in the fine scaly armour of their fish-ancestor (the Branchiosaurs). Some (the Aistopods) were long, snake-like creatures, with shrunken limbs and bodies drawn out until, in some cases, the backbone had 150 vertebrae. They seem to have taken to the thickets, in the growing competition, as the serpents did later, and lost the use of their limbs, which would be merely an encumbrance in winding among the roots and branches. Some (the Microsaurs) were agile little salamander-like organisms, with strong, bony frames and relatively long and useful legs; they look as if they may even have climbed the trees in pursuit of snails and insects. A fourth and more formidable sub-order, the Labyrinthodonts—which take their name from the labyrinthine folds of the enamel in their strong teeth—were commonly several feet in length. Some of them attained a length of seven or eight feet, and had plates of bone over their heads and bellies, while the jaws in their enormous heads were loaded with their strong, labyrinthine teeth. Life on land was becoming as eventful and stimulating as life in the waters.

The general characteristic of these early Amphibia is that they very clearly retain the marks of their fish ancestry. All of them have tails; all of them have either scales or (like many of the fishes) plates of bone protecting the body. In some of the younger specimens the gills can still be clearly traced, but no doubt they were mainly lung-animals. We have seen how the fish obtained its lungs, and need add only that this change in the method of obtaining oxygen for the blood involved certain further changes of a very important nature. Following the fossil record, we do not observe the changes which are taking place in the soft internal organs, but we must not lose sight of them. The heart, for instance, which began as a simple muscular expansion or distension of one of the blood-vessels of some primitive worm, then doubled and became a two-chambered pump in the fish, now develops a partition in the auricle (upper chamber), so that the aerated blood is to some extent separated from the venous blood. This approach toward the warm-blooded type begins in the "mud-fish," and is connected with the development of the lungs. Corresponding changes take place in the arteries, and we shall find that this change in structure is of very great importance in the evolution of the higher types of land-life. The heart of the higher land-animals, we may add, passes through these stages in its embryonic development.

Externally the chief change in the Amphibian is the appearance of definite legs. The broad paddle of the fin is now useless, and its main stem is converted into a jointed, bony limb, with a five-toed foot, spreading into a paddle, at the end. But the legs are still feeble, sprawling supports, letting the heavy body down almost to the ground. The Amphibian is an imperfect, but necessary, stage in evolution. It is an improvement on the Dipneust fish, which now begins to dwindle very considerably in the geological record, but it is itself doomed to give way speedily before one of its more advanced descendants, the Reptile. Probably the giant salamander of modern Japan affords the best suggestion of the large and primitive salamanders of the Coal-forest, while the Caecilia—snake-like Amphibia with scaly skins, which live underground in South America—may not impossibly be degenerate survivors of the curious Aistopods.

Our modern tailless Amphibia, frogs and toads, appear much later in the story of the earth, but they are not without interest here on account of the remarkable capacity which they show to adapt themselves to different surroundings. There are frogs, like the tree-frog of Martinique, and others in regions where water is scarce, which never pass through the tadpole stage; or, to be quite accurate, they lose the gills and tail in the egg, as higher land-animals do. On the other hand, there is a modern Amphibian, the axolotl of Mexico, which retains the gills throughout life, and never lives on land. Dr. Gadow has shown that the lake in which it lives is so rich in food that it has little inducement to leave it for the land. Transferred to a different environment, it may pass to the land, and lose its gills. These adaptations help us to understand the rich variety of Amphibian forms that appeared in the changing conditions of the Carboniferous world.

When we think of the diet of the Amphibia we are reminded of the other prominent representatives of land life at the time. Snails, spiders, and myriapods crept over the ground or along the stalks of the trees, and a vast population of insects filled the air. We find a few stray wings in the Silurian, and a large number of wings and fragments in the Devonian, but it is in the Coal-forest that we find the first great expansion of insect life, with a considerable development of myriapods, spiders, and scorpions. Food was enormously abundant, and the insect at least had no rival in the air, for neither bird nor flying reptile had yet appeared. Hence we find the same generous growth as amongst the Amphibia. Large primitive "may-flies" had wings four or five inches long; great locust-like creatures had fat bodies sometimes twenty inches in length, and soared on wings of remarkable breadth, or crawled on their six long, sprawling legs. More than a thousand species of insects, and nearly a hundred species of spiders and fifty of myriapods, are found in the remains of the Coal-forests.

From the evolutionary point of view these new classes are as obscure in their origin, yet as manifestly undergoing evolution when they do fully appear, as the earlier classes we have considered. All are of a primitive and generalised character; that is to say, characters which are to-day distributed among widely different groups were then concentrated and mingled in one common ancestor, out of which the later groups will develop. All belong to the lowest orders of their class. No Hymenopters (ants, bees, and wasps) or Coleopters (beetles) are found in the Coal-forest; and it will be many millions of years before the graceful butterfly enlivens the landscapes of the earth. The early insects nearly all belong to the lower orders of the Orthopters (cockroaches, crickets, locusts, etc.) and Neuropters (dragon-flies, may-flies, etc.). A few traces of Hemipters (now mainly represented by the degenerate bugs) are found, but nine-tenths of the Carboniferous insects belong to the lowest orders of their class, the Orthopters and Neuropters. In fact, they are such primitive and generalised insects, and so frequently mingle the characteristics of the two orders, that one of the highest authorities, Scudder, groups them in a special and extinct order, the Palmodictyoptera; though this view is not now generally adopted. We shall find the higher orders of insects making their appearance in succession as the story proceeds.

Thus far, then, the insects of the Coal-forest are in entire harmony with the principle of evolution, but when we try to trace their origin and earlier relations our task is beset with difficulties. It goes without saying that such delicate frames as those of the earlier insects had very little chance of being preserved in the rocks until the special conditions of the forest-age set in. We are, therefore, quite prepared to hear that the geologist cannot give us the slenderest information. He finds the wing of what he calls "the primitive bug" (Protocimex), an Hemipterous insect, in the later Ordovician, and the wing of a "primitive cockroach" (Palaeoblattina) in the Silurian. From these we can

141

merely conclude that insects were already numerous and varied. But we have already, in similar difficulties, received assistance from the science of zoology, and we now obtain from that science a most important clue to the evolution of the insect.

In South America, South Africa, and Australasia, which were at one time connected by a great southern continent, we find a little caterpillar-like creature which the zoologist regards with profound interest. It is so curious that he has been obliged to create a special class for it alone—a distinction which will be appreciated when I mention that the neighbouring class of the insects contains more than a quarter of a million living species. This valuable little animal, with its tiny head, round, elongated body, and many pairs of caterpillar-like legs, was until a few decades ago regarded as an Annelid (like the earth-worm). It has, in point of fact, the peculiar kidney-structures (nephridia) and other features of the Annelid, but a closer study discovered in it a character that separated it far from any worm-group. It was found to breathe the air by means of tracheae (little tubes running inward from the surface of the body), as the myriapods, spiders, and insects do. It was, in other words, "a kind of half-way animal between the Arthropods and the Annelids" ("Cambridge Natural History," iv, p. 5), a surviving kink in the lost chain of the ancestry of the insect. Through millions of years it has preserved a primitive frame that really belongs to the Cambrian, if not an earlier, age. It is one of the most instructive "living fossils" in the museum of nature.

Peripatus, as the little animal is called, points very clearly to an Annelid ancestor of all the Tracheates (the myriapods, spiders, and insects), or all the animals that breathe by means of trachere. To understand its significance we must glance once more at an early chapter in the story of life. We saw that a vast and varied wormlike population must have filled the Archaean ocean, and that all the higher lines of animal development start from one or other point in this broad kingdom. The Annelids, in which the body consists of a long series of connected rings or segments, as in the earth-worm, are one of the highest groups of these worm-like creatures, and some branch of them developed a pair of feet (as in the caterpillar) on each segment of the body and a tough, chitinous coat. Thus arose the early Arthropods, on tough-coated, jointed, articulated animals. Some of these remained in the water, breathing by means of gills, and became the Crustacea. Some, however, migrated to the land and developed what we may almost call "lungs"—little tubes entering the body at the skin and branching internally, to bring the air into contact with the blood, the tracheae.

In Peripatus we have a strange survivor of these primitive Annelid-Tracheates of many million years ago. The simple nature of its breathing apparatus suggests that the trachere were developed out of glands in the skin; just as the fish, when it came on land, probably developed lungs from its swimming bladders. The primitive Tracheates, delivered from the increasing carnivores of the waters, grew into a large and varied family, as all such new types do in favourable surroundings. From them in the course of time were evolved the three great classes of the Myriapods (millipedes and centipedes), the Arachnids (scorpions, spiders, and mites), and the Insects. I will not enter into the much-disputed and Obscure question of their nearer relationship. Some derive the Insects from the Myriapods, some the Myriapods from the Insects, and some think they evolved independently; while the rise of the spiders and scorpions is even more obscure.

But how can we see any trace of an Annelid ancestor in the vastly different frames of these animals which are said to descend from it? It is not so difficult as it seems to be at first sight. In the Myriapod we still have the elongated body and successive pairs of legs. In the Arachnid the legs are reduced in number and lengthened, while the various segments of the body are fused in two distinct body-halves, the thorax and the abdomen. In the Insect we have a similar concentration of the primitive long body. The abdomen is composed of a large number (usually nine or ten) of segments which have lost their legs and fused together. In the thorax three segments are still distinctly traceable, with three pairs of legs—now long jointed limbs—as in the caterpillar ancestor; in the Carboniferous insect these three joints in the thorax are particularly clear. In the head four or five segments are fused together. Their limbs have been modified into the jaws or other mouth-appendages, and their separate nerve-centres have combined to form the large ring of nerve-matter round the gullet which represents the brain of the insect.

How, then, do we account for the wings of the insect? Here we can offer nothing more than speculation, but the speculation is not without interest. It may be laid down in principle that the flying animal begins as a leaping animal. The "flying fish" may serve to suggest an early stage in the development of wings; it is a leaping fish, its extended fins merely buoying it, like the surfaces of an aeroplane, and so prolonging its leap away from its pursuer. But the great difficulty is to imagine any part of the smooth-coated primitive insect, apart from the limbs (and the wings of the insect are not developed from legs, like those of the bird), which might have even an initial usefulness in buoying the body as it leaped. It has been suggested, therefore, that the primitive insect returned to the water, as the whale and seal did in the struggle for life of a later period. The fact that the mayfly and dragon-fly spend their youth in the water is thought to confirm this. Returning to the water, the primitive insects would develop gills, like the Crustacea. After a time the stress of life in the water drove them back to the land, and the gills became useless. But the folds or scales of the tough coat, which had covered the gills, would remain as projecting planes, and are thought to have been the rudiment from which a long period of selection evolved the huge wings of the early dragon-flies and mayflies. It is generally believed that the wingless order of insects (Aptera) have not lost, but had never developed, wings, and that the insects with only one or two pairs all descend from an ancestor with three pairs.

The early date of their origin, the delicacy of their structure, and the peculiar form which their larval development has generally assumed, combine to obscure the evolution of the insect, and we must be content for the present with these general indications. The vast unexplored regions of Africa, South America, and Central Australia, may yet yield further clues, and the riddle of insect-metamorphosis may some day betray the secrets which it must hold. For the moment the Carboniferous insects interest us as a rich material for the operation of a coming natural selection. On them, as on all other Carboniferous life, a great trial is about to fall. A very small proportion of them will survive that trial, and they trill be the better organised to maintain themselves and rear their young in the new earth.

The remaining land-life of the Coal-forest is confined to worm-like organisms whose remains are not preserved, and land-snails which do not call for further discussion. We may, in conclusion, glance at the progress of life in the waters. Apart from the appearance of the great fishes and Crustacea, the Carboniferous period was one of great stimulation to aquatic life. Constant changes were taking place in the level and the distribution of land and water. The aspect of our coal seams to-day, alternating between thick layers of sand and mud, shows a remarkable oscillation of the land. Many recent authorities have questioned whether the trees grew on the sites where we find them to-day, and were not rather washed down into the lagoons and shallow waters from higher ground. In that case we could not too readily imagine the forest-clad region sinking below the waves, being buried under the deposits of the rivers, and then emerging, thousands of years later, to receive once more the thick mantle of sombre vegetation. Probably there was less rising and falling of the crust than earlier geologists imagined. But, as one of the most recent and most critical authorities, Professor Chamberlin, observes, the comparative purity of the coal, the fairly uniform thickness of the seams, the bed of clay representing soil at their base, the frequency with which the stumps are still found growing upright (as in the remarkable exposed Coal-forest surface in Glasgow, at the present ground-level), [*] the perfectly preserved fronds and the general mixture of flora, make it highly probable that the coal-seam generally marks the actual site of a Coal-forest, and there were considerable vicissitudes in the distribution of land and water. Great areas of land repeatedly passed beneath the waters, instead of a re-elevation of the land, however, we may suppose that the shallow water was gradually filled with silt and debris from the land, and a fresh forest grew over it.

      * The civic authorities of Glasgow have wisely
exposed and
      protected this instructive piece of Coal-forest in
one of
      their parks. I noticed, however that in the
admirable
      printed information they supply to the public,
they describe
      the trees as "at least several hundred thousand
years old."
      There is no authority in the world who would grant
less than
      ten million years since the Coal-forest period.

These changes are reflected in the progress of marine life, though their influence is probably less than that of the great carnivorous monsters which now fill the waters. The heavy Arthrodirans languish and disappear. The "pavement-toothed" sharks, which at first represent three-fourths of the Elasmobranchs, dwindle in turn, and in the formidable spines which develop on them we may see evidence of the great struggle with the sharp-toothed sharks which are displacing them. The Ostracoderms die out in the presence of these competitors. The smaller fishes (generally Crossopterygii) seem to live mainly in the inland and shore waters, and advance steadily toward the modern types, but none of our modern bony fishes have yet appeared.

More evident still is the effect of the new conditions upon the Crustacea. The Trilobite, once the master of the seas, slowly yields to the stronger competitors, and the latter part of the Carboniferous period sees the last genus of Trilobites finally extinguished. The Eurypterids (large scorpion-like Crustacea, several feet long) suffer equally, and are represented by a few lingering species. The stress favours the development of new and more highly organised Crustacea. One is the Limulus or "king-crab," which seems to be a descendant, or near relative, of the Trilobite, and has survived until modern times. Others announce the coming of the long-tailed Crustacea, of the lobster and shrimp type. They had primitive representatives in the earlier periods, but seem to have been overshadowed by the Trilobites and Eurypterids. As these in turn are crushed, the more highly organised Malacostraca take the lead, and primitive specimens of the shrimp and lobster make their appearance.

The Echinoderms are still mainly represented by the sea-lilies. The rocks which are composed of their remains show that vast areas of the sea-floor must have been covered with groves of sea-lilies, bending on their long, flexible stalks and waving their great flower-like arms in the water to attract food. With them there is now a new experiment in the stalked Echinoderm, the Blastoid, an armless type; but it seems to have been a failure. Sea-urchins are now found in the deposits, and, although their remains are not common, we may conclude that the star-fishes were scattered over the floor of the sea.

For the rest we need only observe that progress and rich diversity of forms characterise the other groups of animals. The Corals now form great reefs, and the finer Corals are gaining upon the coarser. The Foraminifers (the chalk-shelled, one-celled animals) begin to form thick rocks with their dead skeletons; the Radiolaria (the flinty-shelled microbes) are so abundant that more than twenty genera of them have been distinguished in Cornwall and Devonshire. The Brachiopods and Molluscs still abound, but the Molluscs begin to outnumber the lower type of shell-fish. In the Cephalopods we find an increasing complication of the structure of the great spiral-shelled types.

Such is the life of the Carboniferous period. The world rejoices in a tropical luxuriance. Semi-tropical vegetation is found in Spitzbergen and the Antarctic, as well as in North Europe, Asia, and America, and in Australasia; corals and sea-lilies flourish at any part of the earth's surface. Warm, dank, low-lying lands, bathed by warm oceans and steeped in their vapours, are the picture suggested—as we shall see more closely—to the minds of all geologists. In those happy conditions the primitive life of the earth erupts into an abundance and variety that are fitly illustrated in the well-preserved vegetation of the forest. And when the earth has at length flooded its surface with this seething tide of life; when the air is filled with a thousand species of insects, and the forest-floor feels the heavy tread of the giant salamander and the light feet of spiders, scorpions, centipedes, and snails, and the lagoons and shores teem with animals, the Golden Age begins to close, and all the semi-tropical luxuriance is banished. A great doom is pronounced on the swarming life of the Coal-forest period, and from every hundred species of its animals and plants only two or three will survive the searching test.

CHAPTER X. THE PERMIAN REVOLUTION

In an earlier chapter it was stated that the story of life is a story of gradual and continuous advance, with occasional periods of more rapid progress. Hitherto it has been, in these pages, a slow and even advance from one geological age to another, one level of organisation to another. This, it is true, must not be taken too literally. Many a period of rapid change is probably contained, and blurred out of recognition, in that long chronicle of geological events. When a region sinks slowly below the waves, no matter how insensible the subsidence may be, there will often come a time of sudden and vast inundations, as the higher ridges of the coast just dip below the water-level and the lower interior is flooded. When two invading arms of the sea meet at last in the interior of the sinking continent, or when a land-barrier that has for millions of years separated two seas and their populations is obliterated, we have a similar occurrence of sudden and far-reaching change. The whole story of the earth is punctuated with small cataclysms. But we now come to a change so penetrating, so widespread, and so calamitous that, in spite of its slowness, we may venture to call it a revolution.

Indeed, we may say of the remaining story of the earth that it is characterised by three such revolutions, separated by millions of years, which are very largely responsible for the appearance of higher types of life. The facts are very well illustrated by an analogy drawn from the recent and familiar history of Europe.

The socio-political conditions of Europe in the eighteenth century, which were still tainted with feudalism, were changed into the socio-political conditions of the modern world, partly by a slow and continuous evolution, but much more by three revolutionary movements. First there was the great upheaval at the end of the eighteenth century, the tremors of which were felt in the life of every country in Europe. Then, although, as Freeman says, no part of Europe ever returned entirely to its former condition, there was a profound and almost universal reaction. In the 'thirties and 'forties, differing in different countries, a second revolutionary disturbance shook Europe. The reaction after this upheaval was far less severe, and the conditions were permanently changed to a great extent, but a third revolutionary movement followed in the next generation, and from that time the evolution of socio-political conditions has proceeded more evenly.

The story of life on the earth since the Coal-forest period is similarly quickened by three revolutions. The first, at the close of the Carboniferous period, is the subject of this chapter. It is the most drastic and devastating of the three, but its effect, at least on the animal world, will be materially checked by a profound and protracted reaction. At the end of the Chalk period, some millions of years later, there will be a second revolution, and it will have a far more enduring and conspicuous result, though it seem less drastic at the time. Yet there will be something of a reaction after a time, and at length a third revolution will inaugurate the age of man. If it is clearly understood that instead of a century we are contemplating a period of at least ten million years, and instead of a decade of revolution we have a change spread over a hundred thousand years or more, this analogy will serve to convey a most important truth.

The revolutionary agency that broke into the comparatively even chronicle of life near the close of the Carboniferous period, dethroned its older types of organisms, and ushered new types to the lordship of the earth, was cold. The reader will begin to understand why I dwelt on the aspect of the Coal-forest and its surrounding waters. There was, then, a warm, moist earth from pole to pole, not even temporarily chilled and stiffened by a few months of winter, and life spread luxuriantly in the perpetual semi-tropical summer. Then a spell of cold so severe and protracted grips the earth that glaciers glitter on the flanks of Indian and Australian hills, and fields of ice spread over what are now semitropical regions. In some degree the cold penetrates the whole earth. The rich forests shrink slowly into thin tracts of scrubby, poverty-stricken vegetation. The loss of food and the bleak and exacting conditions of the new earth annihilate thousands of species of the older organisms, and the more progressive types are moulded into fitness for the new environment. It is a colossal application of natural selection, and amongst its results are some of great moment.

In various recent works one reads that earlier geologists, led astray by the nebular theory of the earth's origin, probably erred very materially in regard to the climate of primordial times, and that climate has varied less than used to be supposed. It must not be thought that, in speaking of a "Permian revolution," I am ignoring or defying this view of many distinguished geologists. I am taking careful account of it. There is no dispute, however, about the fact that the Permian age witnessed an immense carnage of Carboniferous organisms, and a very considerable modification of those organisms which survived the catastrophe, and that the great agency in this annihilation and transformation was cold. To prevent misunderstanding, nevertheless, it will be useful to explain the controversy about the climate of the earth in past ages which divides modern geologists.

The root of the difference of opinion and the character of the conflicting parties have already been indicated. It is a protest of the "Planetesimalists" against the older, and still general, view of the origin of the earth. As we saw, that view implies that, as the heavier elements penetrated centreward in the condensing nebula, the gases were left as a surrounding shell of atmosphere. It was a mixed mass of gases, chiefly oxygen, hydrogen, nitrogen, and carbon-dioxide (popularly known as "carbonic acid gas"). When the water-vapour settled as ocean on the crust, the atmosphere remained a very dense mixture of oxygen, nitrogen, and carbon-dioxide—to neglect the minor gases. This heavy proportion of carbon-dioxide would cause the atmosphere to act as a glass-house over the surface of the earth, as it does still to some extent. Experiment has shown that an atmosphere containing much vapour and carbon-dioxide lets the heat-rays pass through when they are accompanied by strong light, but checks them when they are separated from the light. In other words, the primitive atmosphere would allow the heat of the sun to penetrate it, and then, as the ground absorbed the light, would retain a large proportion of the heat. Hence the semi-tropical nature of the primitive earth, the moisture, the dense clouds and constant rains that are usually ascribed to it. This condition lasted until the rocks and the forests of the Carboniferous age absorbed enormous quantities of carbon-dioxide, cleared the atmosphere, and prepared an age of chill and dryness such as we find in the Permian.

But the planetesimal hypothesis has no room for this enormous percentage of carbon-dioxide in the primitive atmosphere. Hinc illoe lachrymoe: in plain English, hence the acute quarrel about primitive climate, and the close scanning of the geological chronicle for indications that the earth was not moist and warm until the end of the Carboniferous period. Once more I do not wish to enfeeble the general soundness of this account of the evolution of life by relying on any controverted theory, and we shall find it possible to avoid taking sides.

I have not referred to the climate of the earth in earlier ages, except to mention that there are traces of a local "ice-age" about the middle of the Archaean and the beginning of the Cambrian. As these are many millions of years removed from each other and from the Carboniferous, it is possible that they represent earlier periods more or less corresponding to the Permian. But the early chronicle is so compressed and so imperfectly studied as yet that it is premature to discuss the point. It is, moreover, unnecessary because we know of no life on land in those remote periods, and it is only in connection with life on land that we are interested in changes of climate here. In other words, as far as the present study is concerned, we need only regard the climate of the Devonian and Carboniferous periods. As to this there is no dispute; nor, in fact, about the climate from the Cambrian to the Permian.

As the new school is most brilliantly represented by Professor Chamberlin, [*] it will be enough to quote him. He says of the Cambrian that, apart from the glacial indications in its early part, "the testimony of the fossils, wherever gathered, implies nearly uniform climatic conditions... throughout all the earth wherever records of the Cambrian period are preserved" (ii, 273). Of the Ordovician he says: "All that is known of the life of this era would seem to indicate that the climate was much more uniform than now throughout the areas where the strata of the period are known" (ii, 342). In the Silurian we have "much to suggest uniformity of climate"—in fact, we have just the same evidence for it—and in the Devonian, when land-plants abound and afford better evidence, we find the same climatic equality of living things in the most different latitudes. Finally, "most of the data at hand indicate that the climate of the Lower Carboniferous was essentially uniform, and on the whole both genial and moist" (ii, 518). The "data," we may recall, are in this case enormously abundant, and indicate the climate of the earth from the Arctic regions to the Antarctic. Another recent and critical geologist, Professor Walther ("Geschichte der Erde und des Lebens," 1908), admits that the coal-vegetation shows a uniformly warm climate from Spitzbergen to Africa. Mr. Drew ("The Romance of Modern Geology," 1909) says that "nearly all over the globe the climate was the same—hot, close, moist, muggy" (p. 219).

* An apology is due here in some measure. The work which I
    quote as of Professor Chamberlin ("Geology," 1903) is really
    by two authors, Professors Chamberlin and Salisbury. I
    merely quote Professor Chamberlin for shortness, and because

the particular ideas I refer to are expounded by him in
separate papers. The work is the finest manual in modern
geological literature. I have used it much, in conjunction
with the latest editions of Geikie, Le Conte, and Lupparent,
and such recent manuals as Walther, De Launay, Suess, etc.,
and the geological magazines.

The exception which Professor Chamberlin has in mind when he says "most of the data" is that we find deposits of salt and gypsum in the Silurian and Lower Carboniferous, and these seem to point to the evaporation of lakes in a dry climate. He admits that these indicate, at the most, local areas or periods of dryness in an overwhelmingly moist and warm earth. It is thus not disputed that the climate of the earth was, during a period of at least fifteen million years (from the Cambrian to the Carboniferous), singularly uniform, genial, and moist. During that vast period there is no evidence whatever that the earth was divided into climatic zones, or that the year was divided into seasons. To such an earth was the prolific life of the Coal-forest adapted.

It is, further, not questioned that the temperature of the earth fell in the latter part of the Carboniferous age, and that the cold reached its climax in the Permian. As we turn over the pages of the geological chronicle, an extraordinary change comes over the vegetation of the earth. The great Lepidodendra gradually disappear before the close of the Permian period; the Sigillariae dwindle into a meagre and expiring race; the giant Horsetails (Calamites) shrink, and betray the adverse conditions in their thin, impoverished leaves. New,

157

stunted, hardy trees make their appearance: the Walchia, a tree something like the low Araucarian conifers in the texture of its wood, and the Voltzia, the reputed ancestor of the cypresses. Their narrow, stunted leaves suggest to the imagination the struggle of a handful of pines on a bleak hill-side. The rich fern-population is laid waste. The seed-ferns die out, and a new and hardy type of fern, with compact leaves, the Glossopteris, spreads victoriously over the globe; from Australia it travels northward to Russia, which it reaches in the early Permian, and westward, across the southern continent, to South America. A profoundly destructive influence has fallen on the earth, and converted its rich green forests, in which the mighty Club-mosses had reared their crowns above a sea of waving ferns, into severe and poverty-stricken deserts.

No botanist hesitates to say that it is the coming of a cold, dry climate that has thus changed the face of the earth. The geologist finds more direct evidence. In the Werribee Gorge in Victoria I have seen the marks which Australian geologists have discovered of the ice-age which put an end to their Coal-forests. From Tasmania to Queensland they find traces of the rivers and fields of ice which mark the close of the Carboniferous and beginning of the Permian on the southern continent. In South Africa similar indications are found from the Cape to the Transvaal. Stranger still, the geologists of India have discovered extensive areas of glaciation, belonging to this period, running down into the actual tropics. And the strangest feature of all is that the glaciers of India and Australia flowed, not from the temperate zones toward the tropics, but in the opposite direction. Two great zones of ice-covered land lay north and south of the equator. The total area was probably greater than the enormous area covered with ice in Europe and America during the familiar ice-age of the latest geological period.

Thus the central idea of this chapter, the destructive inroad of a colder climate upon the genial Carboniferous world, is an accepted fact. Critical geologists may suggest that the temperature of the Coal-forest has been exaggerated, and the temperature of the Permian put too low. We are not concerned with the dispute. Whatever the exact change of temperature was, in degrees of the thermometer, it was admittedly sufficient to transform the face of the earth, and bring a mantle of ice over millions of square miles of our tropical and subtropical regions. It remains for us to inquire into the causes of this transformation.

It at once occurs to us that these facts seem to confirm the prevalent idea, that the Coal-forests stripped the air of its carbon-dioxide until the earth shivered in an atmosphere thinner than that of to-day. On reflection, however, it will be seen that, if this were all that happened, we might indeed expect to find enormous ice-fields extending from the poles—which we do not find—but not glaciation in the tropics. Others may think of astronomical theories, and imagine a shrinking or clouding of the sun, or a change in the direction of the earth's axis. But these astronomical theories are now little favoured, either by astronomers or geologists. Professor Lowell bluntly calls them "astrocomic" theories. Geologists think them superfluous. There is another set of facts to be considered in connection with the Permian cold.

As we have seen several times, there are periods when, either owing to the shrinking of the earth or the overloading of the sea-bottoms, or a combination of the two, the land regains its lost territory and emerges from the ocean. Mountain chains rise; new continental surfaces are exposed to the sun and rain. One of the greatest of these upheavals of the land occurs in the latter half of the Carboniferous and the Permian. In the middle of the Carboniferous, when Europe is predominantly a flat, low-lying land, largely submerged, a chain of mountains begins to rise across its central part. From Brittany to the east of Saxony the great ridge runs, and by the end of the Carboniferous it becomes a chain of lofty mountains (of which fragments remain in the Vosges, Black Forest, and Hartz mountains), dragging Central Europe high above the water, and throwing the sea back upon Russia to the north and the Mediterranean region to the south. Then the chain of the Ural Mountains begins to rise on the Russian frontier. By the beginning of the Permian Europe was higher above the water than it had ever yet been; there was only a sea in Russia and a southern sea with narrow arms trailing to the northwest. The continent of North America also had meantime emerged. The rise of the Appalachia and Ouachita mountains completes the emergence of the eastern continent, and throws the sea to the west. The Asiatic continent also is greatly enlarged, and in the southern hemisphere there is a further rise, culminating in the Permian, of the continent ("Gondwana Land") which

160

united South America, South Africa, the Antarctic land, Australia and New Zealand, with an arm to India.

In a word, we have here a physical revolution in the face of the earth. The changes were generally gradual, though they seem in some places to have been rapid and abrupt (Chamberlin); but in summary they amounted to a vast revolution in the environment of animals and plants. The low-lying, swampy, half-submerged continents reared themselves upward from the sea-level, shook the marshes and lagoons from their face, and drained the vast areas that had fostered the growth of the Coal-forests. It is calculated (Chamberlin) that the shallow seas which had covered twenty or thirty million square miles of our continental surfaces in the early Carboniferous were reduced to about five million square miles in the Permian. Geologists believe, in fact, that the area of exposed land was probably greater than it is now.

This lifting and draining of so much land would of itself have a profound influence on life-conditions, and then we must take account of its indirect influence. The moisture of the earlier period was probably due in the main to the large proportion of sea-surface and the absence of high land to condense it. In both respects there is profound alteration, and the atmosphere must have become very much drier. As this vapour had been one of the atmosphere's chief elements for retaining heat at the surface of the earth, the change will involve a great lowering of temperature. The slanting of the raised land would aid this, as, in speeding the rivers, it would promote the circulation of water. Another effect would be to increase the circulation of the atmosphere.

The higher and colder lands would create currents of air that had not been formed before. Lastly, the ocean currents would be profoundly modified; but the effect of this is obscure, and may be disregarded for the moment.

Here, therefore, we have a massive series of causes and effects, all connected with the great emergence of the land, which throw a broad light on the change in the face of the earth. We must add the lessening of the carbon dioxide in the atmosphere. Quite apart from theories of the early atmosphere, this process must have had a great influence, and it is included by Professor Chamberlin among the causes of the world-wide change. The rocks and forests of the Carboniferous period are calculated to have absorbed two hundred times as much carbon as there is in the whole of our atmosphere to-day. Where the carbon came from we may leave open. The Planetesimalists look for its origin mainly in volcanic eruptions, but, though there was much volcanic activity in the later Carboniferous and the Permian, there is little trace of it before the Coal-forests (after the Cambrian). However that may be, there was a considerable lessening of the carbon-dioxide of the atmosphere, and this in turn had most important effects. First, the removal of so much carbon-dioxide and vapour would be a very effective reason for a general fall in the temperature of the earth. The heat received from the sun could now radiate more freely into space. Secondly, it has been shown by experiment that a richness in carbon-dioxide favours Cryptogamous plants (though it is injurious to higher plants), and a reduction of it would therefore be hurtful to the Cryptogams of the Coal-forest. One may almost put it that, in their greed, they exhausted their store. Thirdly, it meant a great purification of the atmosphere, and thus a most important preparation of the earth for higher land animals and plants.

The reader will begin to think that we have sufficiently "explained" the Permian revolution. Far from it. Some of its problems are as yet insoluble. We have given no explanation at all why the ice-sheets, which we would in a general way be prepared to expect, appear in India and Australia, instead of farther north and south. Professor Chamberlin, in a profound study of the period (appendix to vol. ii, "Geology"), suggests that the new land from New Zealand to Antarctica may have diverted the currents (sea and air) up the Indian Ocean, and caused a low atmospheric pressure, much precipitation of moisture, and perpetual canopies of clouds to shield the ice from the sun. Since the outer polar regions themselves had been semi-tropical up to that time, it is very difficult to see how this will account for a freezing temperature in such latitudes as Australia and India. There does not seem to have been any ice at the Poles up to that time, or for ages afterwards, so that currents from the polar regions would be very different from what they are today. If, on the other hand, we may suppose that the rise of "Gondwana Land" (from Brazil to India) was attended by the formation of high mountains in those latitudes, we have the basis, at least, of a more plausible explanation. Professor Chamberlin rejects this supposition on the ground that the traces of ice-action are at or near the sea-level, since we find with them beds containing marine fossils. But this only shows, at the most, that the terminations of the glaciers reached the sea. We know nothing of the height of the land from which they started.

For our main purpose, however, it is fortunately not necessary to clear up these mysteries. It is enough for us that the Carboniferous land rises high above the surface of the ocean over the earth generally. The shallow seas are drained off its surface; its swamps and lagoons generally disappear; its waters run in falling rivers to the ocean. The dense, moist, warm atmosphere that had so long enveloped it is changed into a thinner mantle of gas, through which, night by night, the sun-soaked ground can discharge its heat into space. Cold winds blow over it from the new mountains; probably vast regions of it are swept by icy blasts from the glaciated lands. As these conditions advance in the Permian period, the forests wither and shrink. Of the extraordinarily mixed vegetation which we found in the Coal-forests some few types are fitted to meet the severe conditions. The seed-bearing trees, the thin, needle-leafed trees, the trees with stronger texture of the wood, are slowly singled out by the deepening cold. The golden age of Cryptogams is over. The age of the Cycad and the Conifers is opening. Survivors of the old order linger in the warmer valleys, as one may see to-day tree-ferns lingering in nooks of southern regions while an Antarctic wind is whistling on the hills above them; but over the broad earth the luscious pasturage of the Coal-forest has changed into what is comparatively a cold desert. We must not, of course, imagine too abrupt a change. The earth had

164

been by no means all swamp in the Carboniferous age. The new types were even then developing in the cooler and drier localities. But their hour has come, and there is great devastation among the lower plant population of the earth.

It follows at once that there would be, on land, an equal devastation and a similar selection in the animal world. The vegetarians suffered an appalling reduction of their food; the carnivores would dwindle in the same proportion. Both types, again, would suffer from the enormous changes in their physical surroundings. Vast stretches of marsh, with teeming populations, were drained, and turned into firm, arid plains or bleak hillsides. The area of the Amphibia, for instance, was no less reduced than their food. The cold, in turn, would exercise a most formidable selection. Before the Permian period there was not on the whole earth an animal with a warm-blooded (four-chambered) heart or a warm coat of fur or feathers; nor was there a single animal that gave any further care to the eggs it discharged, and left to the natural warmth of the earth to develop. The extermination of species in the egg alone must have been enormous.

It is impossible to convey any just impression of the carnage which this Permian revolution wrought among the population of the earth. We can but estimate how many species of animals and plants were exterminated, and the reader must dimly imagine the myriads of living things that are comprised in each species. An earlier American geologist, Professor Le Conte, said that not a single Carboniferous species crossed the line of the Permian revolution. This has proved to be an exaggeration, but Professor Chamberlin seems to fall into an exaggeration on the other side when he says that 300 out of 10,000 species survived. There are only about 300 species of animals and plants known in the whole of the Permian rocks (Geikie), and most of these are new. For instance, of the enormous plant-population of the Coal-forests, comprising many thousands of species, only fifty species survived unchanged in the Permian. We may say that, as far as our knowledge goes, of every thirty species of animals and plants in the Carboniferous period, twenty-eight were blotted out of the calendar of life for ever; one survived by undergoing such modifications that it became a new species, and one was found fit to endure the new conditions for a time. We must leave it to the imagination to appreciate the total devastation of individuals entailed in this appalling application of what we call natural selection.

But what higher types of life issued from the womb of nature after so long and painful a travail? The annihilation of the unfit is the seamy side, though the most real side, of natural selection. We ignore it, or extenuate it, and turn rather to consider the advances in organisation by which the survivors were enabled to outlive the great chill and impoverishment.

Unfortunately, if the Permian period is an age of death, it is not an age of burials. The fossil population of its cemeteries is very scanty. Not only is the living population enormously reduced, but the areas that were accustomed to entomb and preserve organisms—the lake and shore deposits—are also greatly reduced. The frames of animals and plants now rot on the dry ground on which they live. Even in the seas, where life must have been much reduced by the general disturbance of conditions, the record is poor. Molluscs and Brachiopods and small fishes fill the list, but are of little instructiveness for us, except that they show a general advance of species. Among the Cephalopods, it is true, we find a notable arrival. On the one hand, a single small straight-shelled Cephalopod lingers for a time with the ancestral form; on the other hand, a new and formidable competitor appears among the coiled-shell Cephalopods. It is the first appearance of the famous Ammonite, but we may defer the description of it until we come to the great age of Ammonites.

Of the insects and their fortunes in the great famine we have no direct knowledge; no insect remains have yet been found in Permian rocks. We shall, however, find them much advanced in the next period, and must conclude that the selection acted very effectively among their thousand Carboniferous species.

The most interesting outcome of the new conditions is the rise and spread of the reptiles. No other sign of the times indicates so clearly the dawn of a new era as the appearance of these primitive, clumsy reptiles, which now begin to oust the Amphibia. The long reign of aquatic life is over; the ensign of progress passes to the land animals. The half-terrestrial, half-aquatic Amphibian deserts the water entirely (in one or more of its branches), and a new and fateful dynasty is founded. Although many of the reptiles will return to the water, when the land sinks once more, the type of the terrestrial quadruped is now fully evolved, and from its early reptilian form will emerge the lords of

167

the air and the lords of the land, the birds and the mammals.

To the uninformed it may seem that no very great advance is made when the reptile is evolved from the Amphibian. In reality the change implies a profound modification of the frame and life of the vertebrate. Partly, we may suppose, on account of the purification of the air, partly on account of the decrease in water surface, the gills are now entirely discarded. The young reptile loses them during its embryonic life—as man and all the mammals and birds do to-day—and issues from the egg a purely lung-breathing creature. A richer blood now courses through the arteries, nourishing the brain and nerves as well as the muscles. The superfluous tissue of the gill-structures is used in the improvement of the ear and mouth-parts; a process that had begun in the Amphibian. The body is raised up higher from the ground, on firmer limbs; the ribs and the shoulder and pelvic bones—the saddles by which the weight of the body is adjusted between the limbs and the backbone—are strengthened and improved. Finally, two important organs for the protection and nurture of the embryo (the amnion and the allantois) make their appearance for the first time in the reptile. In grade of organisation the reptile is really nearer to the bird than it is to the salamander.

Yet these Permian reptiles are so generalised in character and so primitive in structure that they point back unmistakably to an Amphibian ancestry. The actual line of descent is obscure. When the reptiles first appear in the rocks, they are already divided into widely different groups, and must have been evolved some time before. Probably they started from some group or groups of the Amphibia in the later Carboniferous, when, as we saw, the land began to rise considerably. We have not yet recovered, and may never recover, the region where the early forms lived, and therefore cannot trace the development in detail. The fossil archives, we cannot repeat too often, are not a continuous, but a fragmentary, record of the story of life. The task of the evolutionist may be compared to the work of tracing the footsteps of a straying animal across the country. Here and there its traces will be amply registered on patches of softer ground, but for the most part they will be entirely lost on the firmer ground. So it is with the fossil record of life. Only in certain special conditions are the passing forms buried and preserved. In this case we can say only that the Permian reptiles fall into two great groups, and that one of these shows affinities to the small salamander-like Amphibia of the Coal-forest (the Microsaurs), while the other has affinities to the Labyrinthodonts.

A closer examination of these early reptiles may be postponed until we come to speak of the "age of reptiles." We shall see that it is probable that an even higher type of animal, the mammal, was born in the throes of the Permian revolution. But enough has been said in vindication of the phrase which stands at the head of this chapter; and to show how the great Primary age of terrestrial life came to a close. With its new inhabitants the earth enters upon a fresh phase, and thousands of its earlier animals and plants are sealed in their primordial tombs, to await the day when man will break the seals and put flesh once more on the petrified bones.

CHAPTER XI. THE MIDDLE AGES OF THE EARTH

The story of the earth from the beginning of the Cambrian period to the present day was long ago divided by geologists into four great eras. The periods we have already covered—the Cambrian, Ordovician, Silurian, Devonian, Carboniferous, and Permian— form the Primary or Palaeozoic Era, to which the earlier Archaean rocks were prefixed as a barren and less interesting introduction. The stretch of time on which we now enter, at the close of the Permian, is the Secondary or Mesozoic Era. It will be closed by a fresh upheaval of the earth and disturbance of life-conditions in the Chalk period, and followed by a Tertiary Era, in which the earth will approach its modern aspect. At its close there will be another series of upheavals, culminating in a great Ice-age, and the remaining stretch of the earth's story, in which we live, will form the Quaternary Era.

In point of duration these four eras differ enormously from each other. If the first be conceived as comprising sixteen million years—a very moderate estimate—the second will be found to cover less than eight million years, the third less than three million years, and the fourth, the Age of Man, much less than one million years; while the Archaean Age was probably as long as all these put together. But the division is rather based on certain gaps, or "unconformities," in the geological record; and, although the breaches are now partially filled, we saw that they correspond to certain profound and revolutionary disturbances in the face of the earth. We retain them, therefore, as convenient and logical divisions of the biological as well as the geological chronicle, and, instead of passing from one geological period to another, we may, for the rest of the story, take these three eras as wholes, and devote a few chapters to the chief advances made by living things in each era. The Mesozoic Era will be a protracted reaction between two revolutions: a period of low-lying land, great sea-invasions, and genial climate, between two upheavals of the earth. The Tertiary Era will represent a less sharply defined depression, with genial climate and luxuriant life, between two such upheavals.

The Mesozaic ("middle life") Era may very fitly be described as the Middle Ages of life on the earth. It by no means occupies a central position in the chronicle of life from the point of view of time or antiquity, just as the Middle Ages of Europe are by no means the centre of the chronicle of mankind, but its types of animals and plants are singularly transitional between the extinct ancient and the actual modern types. Life has been lifted to a higher level by the Permian revolution. Then, for some millions of years, the sterner process of selection relaxes, the warm bosom of the earth swarms again with a teeming and varied population, and a rich material is provided for the next great application of drastic selective agencies. To a poet it might seem that nature indulges each succeeding and imperfect type of living thing with a golden age before it is dismissed to make place for the higher.

The Mesozoic opens in the middle of the great revolution described in the last chapter. Its first section, the Triassic period, is at first a mere continuation of the Permian. A few hundred species of animals and hardy plants are scattered over a relatively bleak and inhospitable globe. Then the land begins to sink once more. The seas spread in great arms over the revelled continents, the plant world rejoices in the increasing warmth and moisture, and the animals increase in number and variety. We pass into the Jurassic period under conditions of great geniality. Warm seas are found as far north and south as our present polar regions, and the low-lying fertile lands are covered again with rich, if less gigantic, forests, in which hordes of stupendous animals find ample nourishment. The mammal and the bird are already on the stage, but their warm coats and warm blood offer no advantage in that perennial summer, and they await in obscurity the end of the golden age of the reptiles. At the end of the Jurassic the land begins to rise once more. The warm, shallow seas drain off into the deep oceans, and the moist, swampy lands are dried. The emergence continues throughout the Cretaceous (Chalk) period. Chains of vast mountains rise slowly into the air in many parts of the earth, and a new and comparatively rapid change in the vegetation—comparable to that at

172

the close of the Carboniferous—announces the second great revolution. The Mesozoic closes with the dismissal of the great reptiles and the plants on which they fed, and the earth is prepared for its new monarchs, the flowering plants, the birds, and the mammals.

How far this repeated levelling of the land after its repeated upheavals is due to a real sinking of the crust we cannot as yet determine. The geologist of our time is disposed to restrict these mysterious rises and falls of the crust as much as possible. A much more obvious and intelligible agency has to be considered. The vast upheaval of nearly all parts of the land during the Permian period would naturally lead to a far more vigorous scouring of its surface by the rains and rivers. The higher the land, the more effectively it would be worn down. The cooler summits would condense the moisture, and the rains would sweep more energetically down the slopes of the elevated continents. There would thus be a natural process of levelling as long as the land stood out high above the water-line, but it seems probable that there was also a real sinking of the crust. Such subsidences have been known within historic times.

By the end of the Triassic—a period of at least two million years—the sea had reconquered a vast proportion of the territory wrested from it in the Permian revolution. Most of Europe, west of a line drawn from the tip of Norway to the Black Sea, was under water—generally open sea in the south and centre, and inland seas or lagoons in the west. The invasion of the sea continued, and reached its climax, in the Jurassic period. The greater part of Europe was converted into an archipelago. A small continent stood out in the Baltic region. Large areas remained above the sea-level in Austria, Germany, and France. Ireland, Wales, and much of Scotland were intact, and it is probable that a land bridge still connected the west of Europe with the east of America. Europe generally was a large cluster of islands and ridges, of various sizes, in a semi-tropical sea. Southern Asia was similarly revelled, and it is probable that the seas stretched, with little interruption, from the west of Europe to the Pacific. The southern continent had deep wedges of the sea driven into it. India, New Zealand, and Australia were successively detached from it, and by the end of the Mesozoic it was much as we find it to-day. The Arctic continent (north of Europe) was flooded, and there was a great interior sea in the western part of the North American continent.

This summary account of the levelling process which went on during the Triassic and Jurassic will prepare us to expect a return of warm climate and luxurious life, and this the record abundantly evinces. The enormous expansion of the sea—a great authority, Neumayr, believes that it was the greatest extension of the sea that is known in geology—and lowering of the land would of itself tend to produce this condition, and it may be that the very considerable volcanic activity, of which we find evidence in the Permian and Triassic, had discharged great volumes of carbon-dioxide into the atmosphere.

Whatever the causes were, the earth has returned to paradisiacal conditions. The vast ice-fields have gone, the scanty and scrubby vegetation is replaced by luscious forests of cycads, conifers, and ferns, and warmth-loving animals penetrate to what are now the Arctic and Antarctic regions. Greenland and Spitzbergen are fragments of a continent that then bore a luxuriant growth of ferns and cycads, and housed large reptiles that could not now live thousands of miles south of it. England, and a large part of Europe, was a tranquil blue coral-ocean, the fringes of its islands girt with reefs such as we find now only three thousand miles further south, with vast shoals of Ammonites, sometimes of gigantic size, preying upon its living population or evading its monstrous sharks; while the sunlit lands were covered with graceful, palmlike cycads and early yews and pines and cypresses, and quaint forms of reptiles throve on the warm earth or in the ample swamps, or rushed on outstretched wings through the purer air.

It was an evergreen world, a world, apparently, of perpetual summer. No trace is found until the next period of an alternation of summer and winter—no trees that shed their leaves annually, or show annual rings of growth in the wood—and there is little trace of zones of climate as yet. It is true that the sensitive Ammonites differ in the northern and the southern latitudes, but, as Professor Chamberlin says, it is not clear that the difference points to a diversity of climate. We may conclude that the absence of corals higher than the north of England implies a more temperate climate further north, but what Sir A. Geikie calls (with slight exaggeration) "the almost tropical aspect" of Greenland warns us to be cautious. The climate of the mid-Jurassic was very much warmer and more uniform than the climate of the earth to-day. It was an age of great vital expansion. And into this luxuriant world we shall presently find a fresh period of elevation, disturbance, and cold breaking with momentous evolutionary results. Meantime, we may take a closer look at these interesting inhabitants of the Middle Ages of the earth, before they pass away or are driven, in shrunken regiments, into the shelter of the narrowing tropics.

The principal change in the aspect of the earth, as the cold, arid plains and slopes of the Triassic slowly yield the moist and warm ow-lying lands of the Jurassic, to consists in the character of the vegetation. It is wholly intermediate in its forms between that of the primitive forests and that of the modern world. The great Cryptogams of the Carboniferous world—the giant Club-mosses and their kindred—have been slain by the long period of cold and drought. Smaller Horsetails (sometimes of a great size, but generally of the modern type) and Club-mosses remain, but are not a conspicuous feature in the landscape. On the other hand, there is as yet—apart from the Conifers—no trace of the familiar trees and flowers and grasses of the later world. The vast majority of the plants are of the cycad type. These—now confined to tropical and subtropical regions—with the surviving ferns, the new Conifers, and certain trees of the ginkgo type, form the characteristic Mesozoic vegetation.

A few words in the language of the modern botanist will show how this vegetation harmonises with the story of evolution. Plants are broadly divided into the

176

lower kingdom of the Cryptogams (spore-bearing) and the upper kingdom of the Phanerogams (seed-bearing). As we saw, the Primary Era was predominantly the age of Cryptogams; the later periods witness the rise and supremacy of the Phanerogams. But these in turn are broadly divided into a less advanced group, the Gymnosperms, and a more advanced group, the Angiosperms or flowering plants. And, just as the Primary Era is the age of Cryptogams, the Secondary is the age of Gymnosperms, and the Tertiary (and present) is the age of Angiosperms. Of about 180,000 species of plants in nature to-day more than 100,000 are Angiosperms; yet up to the end of the Jurassic not a single true Angiosperm is found in the geological record.

This is a broad manifestation of evolution, but it is not quite an accurate statement, and its inexactness still more strongly confirms the theory of evolution. Though the Primary Era was predominantly the age of Cryptogams, we saw that a very large number of seed-bearing plants, with very mixed characters, appeared before its close. It thus prepares the way for the cycads and conifers and ginkgoes of the Mesozoic, which we may conceive as evolved from one or other branch of the mixed Carboniferous vegetation. We next find that the Mesozoic is by no means purely an age of Gymnosperms. I do not mean merely that the Angiosperms appear in force before its close, and were probably evolved much earlier. The fact is that the Gymnosperms of the Mesozoic are often of a curiously mixed character, and well illustrate the transition to the Angiosperms, though they may not be their actual ancestors. This will be clearer if we glance in succession at the various types of plant which adorned and enriched the Jurassic world.

The European or American landscape—indeed, the aspect of the earth generally, for there are no pronounced zones of climate—is still utterly different from any that we know to-day. No grass carpets the plains; none of the flowers or trees with which we are familiar, except conifers, are found in any region. Ferns grow in great abundance, and have now reached many of the forms with which we are acquainted. Thickets of bracken spread over the plains; clumps of Royal ferns and Hartstongues spring up in moister parts. The trees are conifers, cycads, and trees akin to the ginkgo, or Maidenhair Tree, of modern Japan. Cypresses, yews, firs, and araucarias (the Monkey Puzzle group) grow everywhere, though the species are more primitive than those of today. The broad, fan-like leaves and plum-like fruit of the ginkgoales, of which the temple-gardens of Japan have religiously preserved a solitary descendant, are found in the most distant regions. But the most frequent and characteristic tree of the Jurassic landscape is the cycad.

The cycads—the botanist would say Cycadophyta or Cycadales, to mark them off from the cycads of modern times—formed a third of the whole Jurassic vegetation, while to-day they number only about a hundred species in 180,000, and are confined to warm latitudes. All over the earth, from the Arctic to the Antarctic, their palm-like foliage showered from the top of their generally short stems in the Jurassic. But the most interesting point about them is that a very large branch of them (the Bennettiteae) went far beyond the modern Gymnosperm in their flowers and fruit, and approached the Angiosperms. Their fructifications "rivalled the largest flowers of the present day in structure and modelling" (Scott), and possibly already gave spots of sober colour to the monotonous primitive landscape. On the other hand, they approached the ferns so much more closely than modern cycads do that it is often impossible to say whether Jurassic remains must be classed as ferns or cycads.

We have here, therefore, a most interesting evolutionary group. The botanist finds even more difficulty than the zoologist in drawing up the pedigrees of his plants, but the general features of the larger groups which he finds in succession in the chronicle of the earth point very decisively to evolution. The seed-bearing ferns of the Coal-forest point upward to the later stage, and downward to a common origin with the ordinary spore-bearing ferns. Some of them are "altogether of a cycadean type" (Scott) in respect of the seed. On the other hand, the Bennettiteae of the Jurassic have the mixed characters of ferns, cycads, and flowering plants, and thus, in their turn, point downward to a lower ancestry and upward to the next great stage in plant-development. It is not suggested that the seed-ferns we know evolved into the cycads we know, and these in turn into our flowering plants. It is enough for the student of evolution to see in them so many stages in the evolution of plants up to the Angiosperm level. The gaps between the various groups are less rigid than scientific men used to think.

Taller than the cycads, firmer in the structure of the wood, and destined to survive in thousands of species when the cycads would be reduced to a hundred, were the pines and yews and other conifers of the Jurassic landscape. We saw them first appearing, in the stunted Walchias and Voltzias, during the severe conditions of the Permian period. Like the birds and mammals they await the coming of a fresh period of cold to give them a decided superiority over the cycads. Botanists look for their ancestors in some form related to the Cordaites of the Coal-forest. The ginkgo trees seem to be even more closely related to the Cordaites, and evolved from an early and generalised branch of that group. The Cordaites, we may recall, more or less united in one tree the characters of the conifer (in their wood) and the cycad (in their fruit).

So much for the evolutionary aspect of the Jurassic vegetation in itself. Slender as the connecting links are, it points clearly enough to a selection of higher types during the Permian revolution from the varied mass of the Carboniferous flora, and it offers in turn a singularly varied and rich group from which a fresh selection may choose yet higher types. We turn now to consider the animal population which, directly or indirectly, fed upon it, and grew with its growth. To the reptiles, the birds, and the mammals, we must devote special chapters. Here we may briefly survey the less conspicuous animals of the Mesozoic Epoch.

The insects would be one of the chief classes to benefit by the renewed luxuriance of the vegetation.

The Hymenopters (butterflies) have not yet appeared. They will, naturally, come with the flowers in the next great phase of organic life. But all the other orders of insects are represented, and many of our modern genera are fully evolved. The giant insects of the Coal-forest, with their mixed patriarchal features, have given place to more definite types. Swarms of dragon-flies, may-flies, termites (with wings), crickets, and cockroaches, may be gathered from the preserved remains. The beetles (Coleopters) have come on the scene in the Triassic, and prospered exceedingly. In some strata three-fourths of the insects are beetles, and as we find that many of them are wood-eaters, we are not surprised. Flies (Dipters) and ants (Hymenopters) also are found, and, although it is useless to expect to find the intermediate forms of such frail creatures, the record is of some evolutionary interest. The ants are all winged. Apparently there is as yet none of the remarkable division of labour which we find in the ants to-day, and we may trust that some later period of change may throw light on its origin.

Just as the growth of the forests—for the Mesozoic vegetation has formed immense coal-beds in many parts of the world, even in Yorkshire and Scotland—explains this great development of the insects, they would in their turn supply a rich diet to the smaller land animals and flying animals of the time. We shall see this presently. Let us first glance at the advances among the inhabitants of the seas.

The most important and stimulating event in the seas is the arrival of the Ammonite. One branch of the early shell-fish, it will be remembered, retained the head of its naked ancestor, and lived at the open mouth of its shell, thus giving birth to the Cephalopods. The first form was a long, straight, tapering shell, sometimes several feet long. In the course of time new forms with curved shells appeared, and began to displace the straight-shelled. Then Cephalopods with close-coiled shells, like the nautilus, came, and—such a shell being an obvious advantage—displaced the curved shells. In the Permian, we saw, a new and more advanced type of the coiled-shell animal, the Ammonite, made its appearance, and in the Triassic and Jurassic it becomes the ogre or tyrant of the invertebrate world. Sometimes an inch or less in diameter, it often attained a width of three feet or more across the shell, at the aperture of which would be a monstrous and voracious mouth.

The Ammonites are not merely interesting as extinct monsters of the earth's Middle Ages, and stimulating terrors of the deep to the animals on which they fed. They have an especial interest for the evolutionist. The successive chambers which the animal adds, as it grows, to the habitation of its youth, leave the earlier chambers intact. By removing them in succession in the adult form we find an illustration of the evolution of the elaborate shell of the Jurassic Ammonite. It is an admirable testimony to the validity of the embryonic law we have often quoted—that the young animal is apt to reproduce the past stages of its ancestry—that the order of the building of the shell in the late Ammonite corresponds to the order we trace in its development in the geological chronicle. About a thousand species of Ammonites were developed in the Mesozoic, and none survived the Mesozoic. Like the

Trilobites of the Primary Era, like the contemporary great reptiles on land, the Ammonites were an abortive growth, enjoying their hour of supremacy until sterner conditions bade them depart. The pretty nautilus is the only survivor to-day of the vast Mesozoic population of coiled-shell Cephalopods.

A rival to the Ammonite appeared in the Triassic seas, a formidable forerunner of the cuttle-fish type of Cephalopod. The animal now boldly discards the protecting and confining shell, or spreads over the outside of it, and becomes a "shell-fish" with the shell inside. The octopus of our own time has advanced still further, and become the most powerful of the invertebrates. The Belemnite, as the Mesozoic cuttle-fish is called, attained so large a size that the internal bone, or pen (the part generally preserved), is sometimes two feet in length. The ink-bags of the Belemnite also are sometimes preserved, and we see how it could balk a pursuer by darkening the waters. It was a compensating advantage for the loss of the shell.

In all the other classes of aquatic animals we find corresponding advances. In the remaining Molluscs the higher or more effective types are displacing the older. It is interesting to note that the oyster is fully developed, and has a very large kindred, in the Mesozoic seas. Among the Brachiopods the higher sloping-shoulder type displaces the square-shoulder shells. In the Crustacea the Trilobites and Eurypterids have entirely disappeared; prawns and lobsters abound, and the earliest crab makes its appearance in the English Jurassic rocks. This sudden arrival of a short-tailed Crustacean surprises us less when we learn that the crab has a long tail in its embryonic form, but the actual line of its descent is not clear. Among the Echinoderms we find that the Cystids and Blastoids have gone, and the sea-lilies reach their climax in beauty and organisation, to dwindle and almost disappear in the last part of the Mesozoic. One Jurassic sea-lily was found to have 600,000 distinct ossicles in its petrified frame. The free-moving Echinoderms are now in the ascendant, the sea-urchins being especially abundant. The Corals are, as we saw, extremely abundant, and a higher type (the Hexacoralla) is superseding the earlier and lower (Tetracoralla).

Finally, we find a continuous and conspicuous advance among the fishes. At the close of the Triassic and during the Jurassic they seem to undergo profound and comparatively rapid changes. The reason will, perhaps, be apparent in the next chapter, when we describe the gigantic reptiles which feed on them in the lakes and shore-waters. A greater terror than the shark had appeared in their environment. The Ganoids and Dipneusts dwindle, and give birth to their few modern representatives. The sharks with crushing teeth diminish in number, and the sharp-toothed modern shark attains the supremacy in its class, and evolves into forms far more terrible than any that we know to-day. Skates and rays of a more or less modern type, and ancestral gar-pikes and sturgeons, enter the arena. But the most interesting new departure is the first appearance, in the Jurassic, of bony-framed fishes (Teleosts). Their superiority in organisation soon makes itself felt, and they enter upon the rapid evolution which will, by the next period, give them the first place in the fish world.

Over the whole Mesozoic world, therefore, we find advance and the promise of greater advance. The Permian stress has selected the fittest types to survive from the older order; the Jurassic luxuriance is permitting a fresh and varied expansion of life, in preparation for the next great annihilation of the less fit and selection of the more fit. Life pauses before another leap. The Mesozoic earth—to apply to it the phrase which a geologist has given to its opening phase—welcomes the coming and speeds the parting guest. In the depths of the ocean a new movement is preparing, but we have yet to study the highest forms of Mesozoic life before we come to the Cretaceous disturbances.

CHAPTER XII. THE AGE OF REPTILES

From one point of view the advance of life on the earth seems to proceed not with the even flow of a river, but in the successive waves of an oncoming tide. It is true that we have detected a continuous advance behind all these rising and receding waves, yet their occurrence is a fact of some interest, and not a little speculation has been expended on it. When the great procession of life first emerges out of the darkness of

Archaean times, it deploys into a spreading world of strange Crustaceans, and we have the Age of Trilobites. Later there is the Age of Fishes, then of Cryptogams and Amphibia, and then of Cycads and Reptiles, and there will afterwards be an Age of Birds and Mammals, and finally an Age of Man. But there is no ground for mystic speculation on this circumstance of a group of organisms fording the earth for a few million years, and then perishing or dwindling into insignificance. We shall see that a very plain and substantial process put an end to the Age of the Cycads, Ammonites, and Reptiles, and we have seen how the earlier dynasties ended.

The phrase, however, the Age of Reptiles, is a fitting and true description of the greater part of the Mesozoic Era, which lies, like a fertile valley, between the Permian and the Chalk upheavals. From the bleak heights of the Permian period, or—more probably—from its more sheltered regions, in which they have lingered with the ferns and cycads, the reptiles spread out over the earth, as the summer of the Triassic period advances. In the full warmth and luxuriance of the Jurassic they become the most singular and powerful army that ever trod the earth. They include small lizard-like creatures and monsters more than a hundred feet in length. They swim like whales in the shallow seas; they shrink into the shell of the giant turtle; they rear themselves on towering hind limbs, like colossal kangaroos; they even rise into the air, and fill it with the dragons of the fairy tale. They spread over the whole earth from Australia to the Arctic circle. Then the earth seems to grow impatient of their dominance, and they shrink towards the south, and struggle in a diminished territory. The colossal monsters and the formidable dragons go the way of all primitive life, and a ragged regiment of crocodiles, turtles, and serpents in the tropics, with a swarm of smaller creatures in the fringes of the warm zone, is all that remains, by the Tertiary Era, of the world-conquering army of the Mesozoic reptiles.

They had appeared, as we said, in the Permian period. Probably they had been developed during the later Carboniferous, since we find them already branched into three orders, with many sub-orders, in the Permian. The stimulating and selecting disturbances which culminated in the Permian revolution had begun in the Carboniferous. Their origin is not clear, as the intermediate forms between them and the amphibia are not found. This is not surprising, if we may suppose that some of the amphibia had, in the growing struggle, pushed inland, or that, as the land rose and the waters were drained in certain regions, they had gradually adopted a purely terrestrial life, as some of the frogs have since done. In the absence of water their frames would not be preserved and fossilised. We can, therefore, understand the gap in the record between the amphibia and the reptiles. From their structure we gather that they sprang from at least two different branches of the amphibia. Their remains fall into two great groups, which are known as the Diapsid and the Synapsid reptiles. The former seem to be more closely related to the Microsauria, or small salamander-like amphibia of the Coal-forest; the latter are nearer to the

Labyrinthodonts. It is not suggested that these were their actual ancestors, but that they came from the same early amphibian root.

We find both these groups, in patriarchal forms, in Europe, North America, and South Africa during the Permian period. They are usually moderate in size, but in places they seem to have found good conditions and prospered. A few years ago a Permian bed in Russia yielded a most interesting series of remains of Synapsid reptiles. Some of them were large vegetarian animals, more than twelve feet in length; others were carnivores with very powerful heads and teeth as formidable as those of the tiger. Another branch of the same order lived on the southern continent, Gondwana Land, and has left numerous remains in South Africa. We shall see that they are connected by many authorities with the origin of the mammals. [*] The other branch, the Diapsids, are represented to-day by the curiously primitive lizard of New Zealand, the tuatara (Sphenodon, or Hatteria), of which I have seen specimens, nearly two feet in length, that one did not care to approach too closely. The Diapsids are chiefly interesting, however, as the reputed ancestors of the colossal reptiles of the Jurassic age and the birds.

* These Synapsid reptiles are more commonly known as
  Pareiasauria or Theromorpha.

The purified air of the Permian world favoured the reptiles' being lung-breathers, but the cold would check their expansion for a time. The reptile, it is important to remember' usually leaves its eggs to be hatched by the natural warmth of the ground. But as the cold of the Permian yielded to a genial climate and rich vegetation in the course of the Triassic, the reptiles entered upon their memorable development. The amphibia were now definitely ousted from their position of dominance. The increase of the waters had at first favoured them, and we find more than twenty genera, and some very large individuals, of the amphibia in the Triassic. One of them, the Mastodonsaurus, had a head three feet long and two feet wide. But the spread of the reptiles checked them, and they shrank rapidly into the poor and defenceless tribe which we find them in nature to-day.

To follow the prolific expansion of the reptiles in the semi-tropical conditions of the Jurassic age is a task that even the highest authorities approach with great diffidence. Science is not yet wholly agreed in the classification of the vast numbers of remains which the Mesozoic rocks have yielded, and the affinities of the various groups are very uncertain. We cannot be content, however, merely to throw on the screen, as it were, a few of the more quaint and monstrous types out of the teeming Mesozoic population, and describe their proportions and peculiarities. They fall into natural and intelligible groups or orders, and their features are closely related to the differing regions of the Jurassic world. While, therefore, we must abstain from drawing up settled genealogical trees, we may, as we review in succession the monsters of the land, the waters, and the air, glance at the most recent and substantial conjectures of scientific men as to their origin and connections.

The Deinosaurs (or "terrible reptiles"), the monarchs of the land and the swamps, are the central and outstanding family of the Mesozoic reptiles. As the name implies, this group includes most of the colossal animals, such as the Diplodocus, which the illustrated magazine has made familiar to most people. Fortunately the assiduous research of American geologists and their great skill and patience in restoring the dead forms enable us to form a very fair picture of this family of medieval giants and its remarkable ramifications. [*]

* See, besides the usual authorities, a valuable paper by
  Dr. R. S. Lull, "Dinosaurian Distribution" (1910).

The Diapsid reptiles of the Permian had evolved a group with horny, parrot-like beaks, the Rhyncocephalia (or "beak-headed" reptiles), of which the tuatara of New Zealand is a lingering representative. New Zealand seems to have been cut off from the southern continent at the close of the Permian or beginning of the Triassic, and so preserved for us that very interesting relic of Permian life. From some primitive level of this group, it is generally believed, the great Deinosaurs arose. Two different orders seem to have arisen independently, or diverged rapidly from each other, in different parts of the world. One group seems to have evolved on the "lost Atlantis," the land between Western Europe and America, whence they spread westward to America, eastward over Europe, and southward to the continent which still united Africa and Australia. We find their remains in all these regions. Another stock is believed to have arisen in America.

Both these groups seem to have been more or less biped, rearing themselves on large and powerful hind limbs, and (in some cases, at least) probably using their small front limbs to hold or grasp their food. The first group was carnivorous, the second herbivorous; and, as the reptiles of the first group had four or five toes on each foot, they are known as the Theropods (or "beast-footed"), while those of the second order, which had three toes, are called the Ornithopods (or "bird-footed"). Each of them then gave birth to an order of quadrupeds. In the spreading waters and rich swamps of the later Triassic some of the Theropods were attracted to return to an amphibious life, and became the vast, sprawling, ponderous Sauropods, the giants in a world of giants. On the other hand, a branch of the vegetarian Ornithopods developed heavy armour, for defence against the carnivores, and became, under the burden of its weight, the quadrupedal and monstrous Stegosauria and Ceratopsia. Taking this instructive general view of the spread of the Deinosaurs as the best interpretation of the material we have, we may now glance at each of the orders in succession.

The Theropods varied considerably in size and agility. The Compsognathus was a small, active, rabbit-like creature, standing about two feet high on its hind limbs, while the Megalosaurs stretched to a length of thirty feet, and had huge jaws armed with rows of formidable teeth. The Ceratosaur, a seventeen-foot-long reptile, had hollow bones, and we find this combination of lightness and strength in several members of the group. In many respects the group points more or less significantly toward the birds. The brain is relatively large, the neck long, and the fore limbs might be used for grasping, but had apparently ceased to serve as legs. Many of the Theropods were evidently leaping reptiles, like colossal kangaroos, twenty or more feet in length when they were erect. It is the general belief that the bird began its career as a leaping reptile, and the feathers, or expanded scales, on the front limbs helped at first to increase the leap. Some recent authorities hold, however, that the ancestor of the bird was an arboreal reptile.

To the order of the Sauropods belong most of the monsters whose discovery has attracted general attention in recent years. Feeding on vegetal matter in the luscious swamps, and having their vast bulk lightened by their aquatic life, they soon attained the most formidable proportions. The admirer of the enormous skeleton of Diplodocus (which ran to eighty feet) in the British Museum must wonder how even such massive limbs could sustain the mountain of flesh that must have covered those bones. It probably did not walk so firmly as the skeleton suggests, but sprawled in the swamps or swam like a hippopotamus. But the Diplodocus is neither the largest nor heaviest of its family. The Brontosaur, though only sixty feet long, probably weighed twenty tons. We have its footprints in the rocks to-day, each impression measuring about a square yard. Generally, it is the huge thigh-bones of these monsters that have survived, and give us an idea of their size. The largest living elephant has a femur scarcely four feet long, but the femur of the Atlantosaur measures more than seventy inches, and the femur of the Brachiosaur more than eighty. Many of these Deinosaurs must have measured more than a hundred feet from the tip of the snout to the end of the tail, and stood about thirty feet high from the ground. The European Sauropods did not, apparently, reach the size of their American cousins—so early did the inferiority of Europe begin—but our Ceteosaur seems to have been about fifty feet long and ten feet in height. Its thigh-bone was sixty-four inches long and twenty-seven inches in circumference at the shaft. And in this order of reptiles, it must be remembered, the bones are solid.

To complete the picture of the Sauropods, we must add that the whole class is characterised by the extraordinary smallness of the brain. The twenty-ton Brontosaur had a brain no larger than that of a new-born human infant. Quite commonly the brain of one of these enormous animals is no larger than a man's fist. It is true that, as far as the muscular and sexual labour was concerned, the brain was supplemented by a great enlargement of the spinal cord in the sacral region (at the top of the thighs). This inferior "brain" was from ten to twenty times as large as the brain in the skull. It would, however, be fully occupied with the movement of the monstrous limbs and tail, and the sex-life, and does not add in the least to the "mental" power of the Sauropods. They were stupid, sluggish, unwieldy creatures, swollen parasites upon a luxuriant vegetation, and we shall easily understand their disappearance at the end of the Mesozoic Era, when the age of brawn will yield to an age of brain.

The next order of the Deinosaurs is that of the biped vegetarians, the Ornithopods, which gradually became heavily armoured and quadrupedal. The familiar Iguanodon is the chief representative of this order in

Europe. Walking on its three-toed hind limbs, its head would be fourteen or fifteen feet from the ground. The front part of its jaws was toothless and covered with horn. It had, in fact, a kind of beak, and it also approached the primitive bird in the structure of its pelvis and in having five toes on its small front limbs. Some of the Ornithopods, such as the Laosaur, were small (three or four feet in height) and active, but many of the American specimens attained a great size. The Camptosaur, which was closely related to the Iguanodon in structure, was thirty feet from the snout to the end of the tail, and the head probably stood eighteen feet from the ground. One of the last great representatives of the group in America, the Trachodon, about thirty feet in length, had a most extraordinary head. It was about three and a half feet in length, and had no less than 2000 teeth lining the mouth cavity. It is conjectured that it fed on vegetation containing a large proportion of silica.

In the course of the Jurassic, as we saw, a branch of these biped, bird-footed vegetarians developed heavy armour, and returned to the quadrupedal habit. We find them both in Europe and America, and must suppose that the highway across the North Atlantic still existed.

The Stegosaur is one of the most singular and most familiar representatives of the group in the Jurassic. It ran to a length of thirty feet, and had a row of bony plates, from two to three feet in height, standing up vertically along the ridge of its back, while its tail was armed with formidable spikes. The Scleidosaur, an earlier and smaller (twelve-foot) specimen, also had spines and bony plates to protect it. The Polacanthus and Ankylosaur developed a most effective armour-plating over the rear. As we regard their powerful armour, we seem to see the fierce-toothed Theropods springing from the rear upon the poor-mouthed vegetarians. The carnivores selected the vegetarians, and fitted them to survive. Before the end of the Mesozoic, in fact, the Ornithopods became aggressive as well as armoured. The Triceratops had not only an enormous skull with a great ridged collar round the neck, but a sharp beak, a stout horn on the nose, and two large and sharp horns on the top of the head. We will see something later of the development of horns. The skulls of members of the Ceratops family sometimes measured eight feet from the snout to the ridge of the collar. They were, however, sluggish and stupid monsters, with smaller brains even than the Sauropods.

Such, in broad outline, was the singular and powerful family of the Mesozoic Deinosaurs. Further geological research in all parts of the world will, no doubt, increase our knowledge of them, until we can fully understand them as a great family throwing out special branches to meet the different conditions of the crowded Jurassic age. Even now they afford a most interesting page in the story of evolution, and their total disappearance from the face of the earth in the next geological period will not be unintelligible. We turn from them to the remaining orders of the Jurassic reptiles.

In the popular mind, perhaps, the Ichthyosaur and Plesiosaur are the typical representatives of that extinct race. The two animals, however, belong to very different branches of the reptile world, and are by no means the most formidable of the Mesozoic reptiles. Many orders of the land reptiles sent a branch into the waters in an age which, we saw, was predominantly one of water-surface. The Ichthyosauria ("fish-reptiles") and Thalattosauria ("sea-reptiles") invaded the waters at their first expansion in the later Triassic. The latter groups soon became extinct, but the former continued for some millions of years, and became remarkably adapted to marine life, like the whale at a later period.

The Ichthyosaur of the Jurassic is a remarkably fish-like animal. Its long tapering frame—sometimes forty feet in length, but generally less than half that length—ends in a dip of the vertebral column and an expansion of the flesh into a strong tail-fin. The terminal bones of the limbs depart more and more from the quadruped type, until at last they are merely rows of circular bony plates embedded in the broad paddle into which the limb has been converted. The head is drawn out, sometimes to a length of five feet, and the long narrow jaws are set with two formidable rows of teeth; one specimen has about two hundred teeth. In some genera the teeth degenerate in the course of time, but this merely indicates a change of diet. One fossilised Ichthyosaur of the weaker-toothed variety has been found with the remains of two hundred Belemnites in its stomach. It is a flash of light on the fierce struggle and carnage which some recent writers have vainly striven to attenuate. The eyes, again, which may in the larger animals be fifteen inches in diameter, are protected by a circle of radiating bony plates. In fine, the discovery of young developed skeletons inside the adult frames has taught us that the Ichthgosaur had become viviparous, like the mammal. Cutting its last connection with the land, on which it originated it ceased to lay eggs, and developed the young within its body.

The Ichthyosaur came of the reptile group which we have called the Diapsids. The Plesiosaur seems to belong to the Synapsid branch. In the earlier Mesozoic we find partially aquatic representatives of the line, like the Nothosaur, and in the later Plesiosaur the adaptation to a marine life is complete. The skin has lost its scales, and the front limbs are developed into powerful paddles, sometimes six feet in length. The neck is drawn out until, in some specimens, it is found to consist of seventy-six vertebrae: the longest neck in the animal world. It is now doubted, however, if the neck was very flexible, and, as the jaws were imperfectly joined, the common picture of the Plesiosaur darting its snake-like neck in all directions to seize its prey is probably wrong. It seems to have lived on small food, and been itself a rich diet to the larger carnivores. We find it in all the seas of the Mesozoic world, varying in length from six to forty feet, but it is one of the sluggish and unwieldy forms that are destined to perish in the coming crisis.

The last, and perhaps the most interesting, of the doomed monsters of the Mesozoic was the Pterosaur, or "flying reptile." It is not surprising that in the fierce struggle which is reflected in the arms and armour of the great reptiles, a branch of the family escaped into the upper region. We have seen that there were leaping reptiles with hollow bones, and although the intermediate forms are missing, there is little doubt that the Pterosaur developed from one or more of these leaping Deinosaurs. As it is at first small, when it appears in the early Jurassic—it is disputed in the late Triassic—it probably came from a small and agile Deinosaur, hunted by the carnivores, which relied on its leaping powers for escape. A flapperlike broadening of the fore limbs would help to lengthen the leap, and we must suppose that this membrane increased until the animal could sail through the air, like the flying-fish, and eventually sustain its weight in the air. The wing is, of course, not a feathery frame, as in the bird, but a special skin spreading between the fore limb and the side of the body. In the bat this skin is supported by four elongated fingers of the hand, but in the Pterosaur the fifth (or fourth) finger alone—which is enormously elongated and strengthened—forms its outer frame. It is as if, in flying experiments, a man were to have a web of silk stretching from his arm and an extension of his little finger to the side of his body.

From the small early specimens in the early Jurassic the flying reptiles grow larger and larger until the time of their extinction in the stresses of the Chalk upheaval. Small Pterosaurs continue throughout the period, but from these bat-like creatures we rise until we come to such dragons as the American Pteranodon, with a stretch of twenty-two feet between its extended wings and jaws about four feet long. There were long-tailed Pterosaurs (Ramphorhyncus), sometimes with a rudder-like expansion of the end of the tail, and short-tailed Pterosaurs (Pterodactyl), with compact bodies and keeled breasts, like the bird. In the earlier part of the period they all have the heavy jaws and numerous teeth of the reptile, with four or five well-developed fingers on the front limbs. In the course of time they lose the teeth—an advantage in the distribution of the weight of the body while flying—and develop horny beaks. In the gradual shaping of the breast-bone and head, also, they illustrate the evolution of the bird-form.

But the birds were meantime developing from a quite different stock, and would replace the Pterosaurs at the first change in the environment. There is ground for thinking that these flying reptiles were warm-blooded like the birds. Their hollow bones seem to point to the effective breathing of a warm-blooded animal, and the great vitality they would need in flying points toward the same conclusion. Their brain, too, approached that of the bird, and was much superior to that of the other reptiles. But they had no warm coats to retain their heat, no clavicle to give strength to the wing machinery, and, especially in the later period, they became very weak in the hind limbs (and therefore weak or slow in starting their flight). The coming selection will therefore dismiss them from the scene, with the Deinosaurs and Ammonites, and retain the better organised bird as the lord of the air.

There remain one or two groups of the Mesozoic reptiles which are still represented in nature. The turtle-group (Chelonia) makes its appearance in the Triassic and thrives in the Jurassic. Its members are extinct and primitive forms of the thick-shelled reptiles, but true turtles, both of marine and fresh water, abound before the close of the Mesozoic. The sea-turtles attain an enormous size. Archelon, one of the primitive types, measured about twelve feet across the shell. Another was thirteen feet long and fifteen feet from one outstretched flipper to the other. In the Chalk period they form more than a third of the reptile remains in some regions. They are extremely interesting in that they show, to some extent, the evolution of their characteristic shell. In some of the larger specimens the ribs have not yet entirely coalesced.

The Crocodilians also appear in the later Triassic, abound in the Jurassic, and give way before the later types, the true Crocodiles, in the Cretaceous. They were marine animals with naked skin, a head and neck something like that of the Ichthyosaur, and paddles like those of the Plesiosaur. Their back limbs, however, were not much changed after their adaptation to life in the sea, and it is concluded that they visited the land to lay their eggs. The Teleosaur was a formidable narrow-spouted reptile, somewhat resembling the crocodiles of the Ganges in the external form of the jaws. The modern crocodiles, which replaced this ancient race of sea-crocodiles, have a great advantage over them in the fact that their nostrils open into the mouth in its lower depths. They can therefore close their teeth on their prey under water and breathe through the nose.

Snakes are not found until the close of the Mesozoic, and do not figure in its characteristic reptile population. We will consider them later. But there was a large group of reptiles in the later Mesozoic seas which more or less correspond to the legendary idea of a sea-serpent. These Dolichosaurs ("long reptiles") appear at the beginning of the Chalk period, and develop into a group, the Mososaurians, which must have added considerably to the terrors of the shore-waters. Their slender scale-covered bodies were commonly twenty to thirty feet in length. The supreme representative of the order, the Mososaur, of which about forty species are known, was sometimes seventy-five feet long. It had two pairs of paddles—so that the name of sea-serpent is very imperfectly applicable—and four rows of formidable teeth on the roof of its mouth. Like the Deinosaurs and Pterosaurs, the order was doomed to be entirely extinguished after a brief supremacy in its environment.

From this short and summary catalogue the reader will be able to form some conception of the living inhabitants of the Mesozoic world. It is assuredly the Age of Reptiles. Worms, snails, and spiders were, we may assume, abundant enough, and a great variety of insects flitted from tree to tree or sheltered in the fern brakes. But the characteristic life, in water and on land, was the vast and diversified family of the reptiles. In the western and the eastern continent, and along the narrowing bridge that still united them, in the northern hemisphere and the southern, and along every ridge of land that connected them, these sluggish but formidable monsters filled the stage. Every conceivable device in the way of arms and armour, brute strength and means of escape, seemed to be adopted in their development, as if they were the final and indestructible outcome of the life-principle. And within a single geological period the overwhelming

201

majority of them, especially the larger and more formidable of them, were ruthlessly slain, leaving not a single descendant on the earth. Let us see what types of animals were thus preferred to them in the next great application of selective processes.

CHAPTER XIII. THE BIRD AND THE MAMMAL

In one of his finest stories, Sur La Pierre Blanche, Anatole France has imagined a group of Roman patricians discussing the future of their Empire. The Christians, who are about to rise to power on their ruin, they dismiss with amiable indifference as one of the little passing eccentricities of the religious life of their time. They have not the dimmest prevision, even as the dream of a possibility, that in a century or two the Empire of Rome will lie in the dust, and the cross will tower above all its cities from York to Jerusalem. If we might for a moment endow the animals of the Mesozoic world with AEsopian wisdom, we could imagine some such discussion taking place between a group of Deinosaur patricians. They would reflect with pride on the unshakable empire of the reptiles, and perhaps glance with disdain at two types of animals which hid in the recesses or fled to the hills of the Jurassic world. And before another era of the earth's story opened, the reptile race would be dethroned, and these hunted and despised and feeble eccentricities of Mesozoic life would become the masters of the globe.

These two types of organisms were the bird and the mammal. Both existed in the Jurassic, and the mammals at least had many representatives in the Triassic. In other words, they existed, with all their higher organisation, during several million years without attaining power. The mammals remained, during at least 3,000,000 years, a small and obscure caste, immensely overshadowed by the small-brained reptiles. The birds, while making more progress, apparently, than the mammals, were far outnumbered by the flying reptiles until the last part of the Mesozoic. Then there was another momentous turn of the wheel of fate, and they emerged from their obscurity to assume the lordship of the globe.

In earlier years, when some serious hesitation was felt by many to accept the new doctrine of evolution, a grave difficulty was found in the circumstance that new types—not merely new species and new genera, but new orders and even sub-classes—appeared in the geological record quite suddenly. Was it not a singular coincidence that in ALL cases the intermediate organisms between one type and another should have wholly escaped preservation? The difficulty was generally due to an imperfect acquaintance with the conditions of the problem. The fossil population of a period is only that fraction of its living population which happened to be buried in a certain kind of deposit under water of a certain depth. We shall read later of insects being preserved in resin (amber), and we have animals (and even bacteria) preserved in trees from the Coal-forests. Generally speaking, however, the earth has buried only a very minute fraction of its land-population. Moreover, only a fraction of the earth's cemeteries have yet been opened. When we further reflect that the new type of organism, when it first appears, is a small and local group, we see what the chances are of our finding specimens of it in a few scattered pages of a very fragmentary record of the earth's life. We shall see that we have discovered only about ten skeletons or fragments of skeletons of the men who lived on the earth before the Neolithic period; a stretch of some hundreds of thousands of years, recorded in the upper strata of the earth.

Whatever serious difficulty there ever was in this scantiness of intermediate types is amply met by the fact that every fresh decade of search in the geological tombs brings some to light. We have seen many instances of this—the seed-bearing ferns and flower-bearing cycads, for example, found in the last decade—and will see others. But one of the most remarkable cases of the kind now claims our attention. The bird was probably evolved in the late Triassic or early Jurassic. It appears in abundance, divided into several genera, in the Chalk period. Luckily, two bird-skeletons have been found in the intermediate period, the Jurassic, and they are of the intermediate type, between the reptile and the bird, which the theory of evolution would suggest. But for the fortunate accident of these two birds being embedded in an ancient Bavarian mud-layer, which happened to be opened, for commercial purposes, in the second half of the nineteenth century, critics of evolution—if there still were any in the world of science—might be repeating to-day that the transition from the reptile to the bird was unthinkable in theory and unproven in fact.

The features of the Archaeopteryx ("primitive bird") have been described so often, and such excellent pictorial restorations of its appearance may now be seen, that we may deal with it briefly. We have in it a most instructive combination of the characters of the bird and the reptile. The feathers alone, the imprint of which is excellently preserved in the fine limestone, would indicate its bird nature, but other anatomical distinctions are clearly seen in it. "There is," says Dr. Woodward, "a typical bird's 'merrythought' between the wings, and the hind leg is exactly that of a perching bird." In other words, it has the shoulder-girdle and four-toed foot, as well as the feathers, of a bird. On the other hand, it has a long tail (instead of a terminal tuft of feathers as in the bird) consisting of twenty-one vertebrae, with the feathers springing in pairs from either side; it has biconcave vertebrae, like the fishes, amphibia, and reptiles; it has teeth in its jaws; and it has three complete fingers, free and clawed, on its front limbs.

As in the living Peripatus, therefore, we have here a very valuable connecting link between two very different types of organisms. It is clear that one of the smaller reptiles—the Archaeopteryx is between a pigeon and a crow in size—of the Triassic period was the ancestor of the birds. Its most conspicuous distinction was that it developed a coat of feathers. A more important difference between the bird and the reptile is that the heart of the bird is completely divided into four chambers, but, as we saw, this probably occurred also in the other flying reptiles. It may be said to be almost a condition of the greater energy of a flying animal. When the heart has four complete chambers, the carbonised blood from the tissues of the body can be conveyed direct to the lungs for purification, and the aerated blood taken direct to the tissues, without any mingling of the two. In the mud-fish and amphibian, we saw, the heart has two chambers (auricles) above, but one (ventricle) below, in which the pure and impure blood mingle. In the reptiles a partition begins to form in the lower chamber. In the turtle it is so nearly complete that the venous and the arterial blood are fairly separated; in the crocodile it is quite complete, though the arteries are imperfectly arranged. Thus the four-chambered heart of the bird and mammal is not a sudden and inexplicable development. Its advantage is enormous in a cold climate. The purer supply of blood increases the combustion in the tissues, and the animal maintains its temperature and vitality when the surrounding air falls in temperature. It ceases to be "cold-blooded."

But the bird secures a further advantage, and here it outstrips the flying reptile. The naked skin of the Pterosaur would allow the heat to escape so freely when the atmosphere cooled that a great strain would be laid on its vitality. A man lessens the demand on his vitality in cold regions by wearing clothing. The bird somehow obtained clothing, in the shape of a coat of feathers, and had more vitality to spare for life-purposes in a falling temperature. The reptile is strictly limited to one region, the bird can pass from region to region as food becomes scarce.

The question of the origin of the feathers can be discussed only from the speculative point of view, as they are fully developed in the Archaeopteryx, and there is no approach toward them in any other living or fossil organism. But a long discussion of the problem has convinced scientific men that the feathers are evolved from the scales of the reptile ancestor. The analogy between the shedding of the coat in a snake and the moulting of a bird is not uninstructive. In both cases the outer skin or epidermis is shedding an old growth, to be replaced by a new one. The covering or horny part of the scale and the feather are alike growths from the epidermis, and the initial stages of the growth have certain analogies. But beyond this general conviction that the feather is a development of the scale, we cannot proceed with any confidence. Nor need we linger in attempting to trace the gradual modification of the skeleton, owing to the material change in habits. The horny beak and the reduction of the toes are features we have already encountered in the reptile, and the modification of the pelvis, breast-bone, and clavicle are a natural outcome of flight.

In the Chalk period we find a large number of bird remains, of about thirty different species, and in some respects they resume the story of the evolution of the bird. They are widely removed from our modern types of birds, and still have teeth in the jaws. They are of two leading types, of which the Ichthyornis and Hesperornis are the standard specimens. The Ichthyornis was a small, tern-like bird with the power of flight strongly developed, as we may gather from the frame of its wings and the keel-shaped structure of its breast-bone. Its legs and feet were small and slender, and its long, slender jaws had about twenty teeth on each side at the bottom. No modern bird has teeth; though the fact that in some modern species we find the teeth appearing in a rudimentary form is another illustration of the law that animals tend to reproduce ancestral features in their development. A more reptilian character in the Ichthyornis group is the fact that, unlike any modern bird, but like their reptile ancestors, they had biconcave vertebrae. The brain was relatively poor. We are still dealing with a type intermediate in some respects between the reptile and the modern bird. The gannets, cormorants, and pelicans are believed to descend from some branch of this group.

The other group of Cretaceous birds, of the Hesperornis type, show an actual degeneration of the power of flight through adaptation to an environment in which it was not needed, as happened, later, in the kiwi of New Zealand, and is happening in the case of the barn-yard fowl. These birds had become divers. Their wings had shrunk into an abortive bone, while their powerful legs had been peculiarly fitted for diving. They stood out at right angles to the body, and seem to have developed paddles. The whole frame suggests that the bird could neither walk nor fly, but was an excellent diver and swimmer. Not infrequently as large as an ostrich (five to six feet high), with teeth set in grooves in its jaws, and the jaws themselves joined as in the snake, with a great capacity of bolting its prey, the Hesperornis would become an important element in the life of the fishes. The wing-fingers have gone, and the tail is much shortened, but the grooved teeth and loosely jointed jaws still point back to a reptilian ancestry.

These are the only remains of bird-life that we find in the Mesozoic rocks. Admirably as they illustrate the evolution of the bird from the reptile, they seem to represent a relatively poor development and spread of one of the most advanced organisms of the time. It must be understood that, as we shall see, the latter part of the Chalk period does not belong to the depression, the age of genial climate, which I call the Middle Ages of the earth, but to the revolutionary period which closes it. We may say that the bird, for all its advances in organisation, remains obscure and unprosperous as long as the Age of Reptiles lasts. It awaits the next massive uplift of the land and lowering of temperature.

In an earlier chapter I hinted that the bird and the mammal may have been the supreme outcomes of the series of disturbances which closed the Primary Epoch and devastated its primitive population. As far as the bird is concerned, this may be doubted on the ground that it first appears in the upper or later Jurassic, and is even then still largely reptilian in character. We must remember, however, that the elevation of the land and the cold climate lasted until the second part of the Triassic, and it is generally agreed that the bird may have been evolved in the Triassic. Its slow progress after that date is not difficult to understand. The advantage of a four-chambered heart and warm coat would be greatly reduced when the climate became warmer. The stimulus to advance would relax. The change from a coat of scales to a coat of feathers obviously means adaptation to a low temperature, and there is nothing to prevent us from locating it in the Triassic, and indeed no later known period of cold in which to place it.

It is much clearer that the mammals were a product of the Permian revolution. They not only abound throughout the Jurassic, in which they are distributed in more than thirty genera, but they may be traced into the Triassic itself. Both in North America and Europe we find the teeth and fragments of the jaws of small animals which are generally recognised as mammals. We cannot, of course, from a few bones deduce that there already, in the Triassic, existed an animal with a fully developed coat of fur and an apparatus, however crude, in the breast for suckling the young. But these bones so closely resemble the bones of the lowest mammals of to-day that this seems highly probable. In the latter part of the long period of cold it seems that some reptile exchanged its scales for tufts of hair, developed a four-chambered heart, and began the practice of nourishing the young from its own blood which would give the mammals so great an ascendancy in a colder world.

Nor can we complain of any lack of evidence connecting the mammal with a reptile ancestor. The earliest remains we find are of such a nature that the highest authorities are still at variance as to whether they should be classed as reptilian or mammalian. A skull and a fore limb from the Triassic of South Africa (Tritylodon and Theriodesmus) are in this predicament.

It will be remembered that we divided the primitive reptiles of the Permian period into two great groups, the Diapsids and Synapsids (or Theromorphs). The former group have spread into the great reptiles of the Jurassic; the latter have remained in comparative obscurity. One branch of these Theromorph reptiles approach the mammals so closely in the formation of the teeth that they have received the name "of the Theriodonts", or "beast-toothed" reptiles. Their teeth are, like those of the mammals, divided into incisors, canines (sometimes several inches long), and molars; and the molars have in some cases developed cusps or tubercles. As the earlier remains of mammals which we find are generally teeth and jaws, the resemblance of the two groups leads to some confusion in classifying them, but from our point of view it is not unwelcome. It narrows the supposed gulf between the reptile and the mammal, and suggests very forcibly the particular branch of the reptiles to which we may look for the ancestry of the mammals. We cannot say that these Theriodont reptiles were the ancestors of the mammals. But we may conclude with some confidence that they bring us near to the point of origin, and probably had at least a common ancestor with the mammals.

The distribution of the Theriodonts suggests a further idea of interest in regard to the origin of the mammals. It would be improper to press this view in the present state of our knowledge, yet it offers a plausible theory of the origin of the mammals. The Theriodonts seem to have been generally confined to the southern continent, Gondwana Land (Brazil to

Australia), of which an area survives in South Africa. It is there also that we find the early disputed remains of mammals. Now we saw that, during the Permian, Gondwana Land was heavily coated with ice, and it seems natural to suppose that the severe cold which the glacial fields would give to the whole southern continent was the great agency in the evolution of the highest type of the animal world. From this southern land the new-born mammals spread northward and eastward with great rapidity. Fitted as they were to withstand the rigorous conditions which held the reptiles and amphibia in check, they seemed destined to attain at once the domination of the earth. Then, as we saw, the land was revelled once more until its surface broke into a fresh semi-tropical luxuriance, and the Deinosaurs advanced to their triumph. The mammals shrank into a meagre and insignificant population, a scattered tribe of small insect-eating animals, awaiting a fresh refrigeration of the globe.

The remains of these interesting early mammals, restricted, as they generally are, to jaws and teeth and a few other bones that cannot in themselves be too confidently distinguished from those of certain reptiles, may seem insufficient to enable us to form a picture of their living forms. In this, however, we receive a singular and fortunate assistance. Some of them are found living in nature to-day, and their distinctly reptilian features would, even if no fossil remains were in existence, convince us of the evolution of the mammals.

The southern continent on which we suppose the mammals to have originated had its eastern termination in Australia. New Zealand seems to have been detached early in the Mesozoic, and was never reached by the mammals. Tasmania was still part of the Australian continent. To this extreme east of the southern continent the early mammals spread, and then, during either the Jurassic or the Cretaceous, the sea completed its inroad, and severed Australia permanently from the rest of the earth. The obvious result of this was to shelter the primitive life of Australia from invasion by higher types, especially from the great carnivorous mammals which would presently develop. Australia became, in other words, a "protected area," in which primitive types of life were preserved from destruction, and were at the same time sheltered from those stimulating agencies which compelled the rest of the world to advance. "Advance Australia" is the fitting motto of the present human inhabitants of that promising country; but the standard of progress has been set up in a land which had remained during millions of years the Chinese Empire of the living world. Australia is a fragment of the Middle Ages of the earth, a province fenced round by nature at least three million years ago and preserving, amongst its many invaluable types of life, representatives of that primitive mammal population which we are seeking to understand.

It is now well known that the Duckbill or Platypus (Ornithorhyncus) and the Spiny Anteater (Echidna) of Australia and Tasmania—with one representative of the latter in New Guinea, which seems to have been still connected—are semi-reptilian survivors of the first animals to suckle their young. Like the reptiles they lay tough-coated eggs and have a single outlet for the excreta, and they have a reptilian arrangement of the bones of the shoulder-girdle; like the mammals, they have a coat of hair and a four-chambered heart, and they suckle the young. Even in their mammalian features they are, as the careful research of Australian zoologists has shown, of a transitional type. They are warm-blooded, but their temperature is much lower than that of other mammals, and varies appreciably with the temperature of their surroundings. [*] Their apparatus for suckling the young is primitive. There are no teats, and the milk is forced by the mother through simple channels upon the breast, from which it is licked by the young. The Anteater develops her eggs in a pouch. They illustrate a very early stage in the development of a mammal from a reptile; and one is almost tempted to see in their timorous burrowing habits a reminiscence of the impotence of the early mammals after their premature appearance in the Triassic.

* See Lucas and Le Soulf's Animals of Australia, 1909.

The next level of mammal life, the highest level that it attains in Australia (apart from recent invasions), is the Marsupial. The pouched animals (kangaroo, wallaby, etc.) are the princes of pre-human life in Australia, and represent the highest point that life had reached when that continent was cut off from the rest of the world. A few words on the real significance of the pouch, from which they derive their name, will suffice to explain their position in the story of evolution.

Among the reptiles the task of the mother ends, as a rule, with the laying of the egg. One or two modern reptiles hatch the eggs, or show some concern for them, but the characteristic of the reptile is to discharge its eggs upon the warm earth and trouble no further about its young. It is a reminiscence of the warm primitive earth. The bird and mammal, born of the cooling of the earth, exhibit the beginning of that link between mother and offspring which will prove so important an element in the higher and later life of the globe. The bird assists the development of the eggs with the heat of her own body, and feeds the young. The mammal develops the young within the body, and then feeds them at the breast.

But there is a gradual advance in this process. The Duckbill lays its eggs just like the reptile, but provides a warm nest for them at the bottom of its burrow. The Anteater develops a temporary pouch in its body, when it lays an egg, and hatches the egg in it. The Marsupial retains the egg in its womb until the young is advanced in development, then transfers the young to the pouch, and forces milk into its mouth from its breasts. The real reason for this is that the Marsupial falls far short of the higher mammals in the structure of the womb, and cannot fully develop its young therein. It has no placenta, or arrangement by which the blood-vessels of the mother are brought into connection with the blood-vessels of the foetus, in order to supply it with food until it is fully developed. The Marsupial, in fact, only rises above the reptile in hatching the egg within its own body, and then suckling the young at the breast.

These primitive mammals help us to reconstruct the mammal life of the Mesozoic Epoch. The bones that we have are variously described in geological manuals as the remains of Monotremes, Marsupials, and Insectivores. Many of them, if not most, were no doubt insect-eating animals, but there is no ground for supposing that what are technically known as Insectivores (moles and shrews) existed in the Mesozoic. On the other hand, the lower jaw of the Marsupial is characterised by a peculiar hooklike process, and this is commonly found in Mesozoic jaws. This circumstance, and the witness of Australia, permit us, perhaps, to regard the Jurassic mammals as predominantly marsupial. It is more difficult to identify Monotreme remains, but the fact that Monotremes have survived to this day in Australia, and the resemblance of some of the Mesozoic teeth to those found for a time in the young Duckbill justify us in assuming that a part of the Mesozoic mammals correspond to the modern Monotremes. Not single specimen of any higher, or placental, mammal has yet been found in the whole Mesozoic Era.

We must, however, beware of simply transferring to the Mesozoic world the kinds of Monotremes and Marsupials which we know in nature to-day. In some of the excellent "restorations" of Mesozoic life which are found in recent illustrated literature the early mammal is represented with an external appearance like that of the Duckbill. This is an error, as the Duckbill has been greatly modified in its extremities and mouth-parts by its aquatic and burrowing habits. As we have no complete skeletons of these early mammals we must abstain from picturing their external appearance. It is enough that the living Monotreme and Marsupial so finely illustrate the transition from a reptilian to a mammalian form. There may have been types more primitive than the Duckbill, and others between the Duckbill and the Marsupial. It seems clear, at least, that two main branches, the Monotremes and Marsupials, arose from the primitive mammalian root. Whether either of these became in turn the parent of the higher mammals we will inquire later. We must first consider the fresh series of terrestrial disturbances which, like some gigantic sieve, weeded out the grosser types of organisms, and cleared the earth for a rapid and remarkable expansion of these primitive birds and mammals.

We have attended only to a few prominent characters in tracing the line of evolution, but it will be understood that an advance in many organs of the body is implied in these changes. In the lower mammals the diaphragm, or complete partition between the organs of the breast and those of the abdomen, is developed. It is not a sudden and mysterious growth, and its development in the embryo to-day corresponds to the suggestion of its development which the zoologist gathers from the animal series. The ear also is now fully developed. How far the fish has a sense of hearing is not yet fully determined, but the amphibian certainly has an organ for the perception of waves of sound. Parts of the discarded gill-arches are gradually transformed into the three bones of the mammal's internal ear; just as other parts are converted into mouth cartilages, and as—it is believed—one of the gill clefts is converted into the Eustachian tube. In the Monotreme and Marsupial the ear-hole begins to be covered with a shell of cartilage; we have the beginning of the external ear. The jaws, which are first developed in the fish, now articulate more perfectly with the skull. Fat-glands appear in the skin, and it is probably from a group of these that the milk-glands are developed. The origin of the hairs is somewhat obscure. They are not thought to be, like the bird's feathers, modifications of the reptile's scales, but to have been evolved from other structures in the skin, possibly under the protection of the scales.

My purpose is, however, rather to indicate the general causes of the onward advance of life than to study organs in detail—a vast subject—or construct pedigrees. We therefore pass on to consider the next great stride that is taken by the advancing life of the earth. Millions of years of genial climate and rich vegetation have filled the earth with a prolific and enormously varied population. Over this population the hand of natural selection is outstretched, as it were, and we are about to witness another gigantic removal of older types of life and promotion of those which contain the germs of further advance. As we have already explained, natural selection is by no means inactive during these intervening periods of warmth. We have seen the ammonites and reptiles, and even the birds and mammals, evolve into hundreds of species during the Jurassic period. The constant evolution of more effective types of carnivores and their spread into new regions, the continuous changes in the distribution of land and water, the struggle for food in a growing population, and a dozen other causes, are ever at work. But the great and comprehensive changes in the face of the earth which close the eras of the geologist seem to give a deeper and quicker stimulus to its population and result in periods of especially rapid evolution. Such a change now closes the Mesozoic Era, and inaugurates the age of flowering plants, of birds, and of mammals.

CHAPTER XIV. IN THE DAYS OF THE CHALK

In accordance with the view of the later story of the earth which was expressed on an earlier page, we now come to the second of the three great revolutions which have quickened the pulse of life on the earth. Many men of science resent the use of the word revolution, and it is not without some danger. It was once thought that the earth was really shaken at times by vast and sudden cataclysms, which destroyed its entire living population, so that new kingdoms of plants and animals had to be created. But we have interpreted the word revolution in a very different sense. The series of changes and disturbances to which we give the name extended over a period of hundreds of thousands of years, and they were themselves, in some sense, the creators of new types of organisms. Yet they are periods that stand out peculiarly in the comparatively even chronicle of the earth. The Permian period transformed the face of the earth; it lifted the low-lying land into a massive relief, drew mantles of ice over millions of miles of its surface, set volcanoes belching out fire and fumes in many parts, stripped it of its great forests, and slew the overwhelming majority of its animals. On the scale of geological time it may be called a revolution.

It must be confessed that the series of disturbances which close the Secondary and inaugurate the Tertiary Era cannot so conveniently be summed up in a single formula. They begin long before the end of the Mesozoic, and they continue far into the Tertiary, with intervals of ease and tranquillity. There seems to have been no culminating point in the series when the uplifted earth shivered in a mantle of ice and snow. Yet I propose to retain for this period—beginning early in the Cretaceous (Chalk) period and extending into the Tertiary—the name of the Cretaceous Revolution. I drew a fanciful parallel between the three revolutions which have quickened the earth since the sluggish days of the Coal-forest and the three revolutionary movements which have changed the life of modern Europe. It will be remembered that, whereas the first of these European revolutions was a sharp and massive upheaval, the second consisted in a more scattered and irregular series of disturbances, spread over the fourth and fifth decades of the nineteenth century; but they amounted, in effect, to a revolution.

So it is with the Cretaceous Revolution. In effect it corresponds very closely to the Permian Revolution. On the physical side it includes a very considerable rise of the land over the greater part of the globe, and the formation of lofty chains of mountains; on the botanical side it means the reduction of the rich Mesozoic flora to a relatively insignificant population, and the appearance and triumphant spread of the flowering plants, on the zoological side it witnesses the complete extinction of the Ammonites, Deinosaurs, and Pterosaurs, an immense reduction of the reptile world generally, and a victorious expansion of the higher insects, birds, and mammals; on the climatic side it provides the first definite evidence of cold zones of the earth and cold seasons of the year, and seems to represent a long, if irregular, period of comparative cold. Except, to some extent, the last of these points, there is no difference of opinion, and therefore, from the evolutionary point of view, the Cretaceous period merits the title of a revolution. All these things were done before the Tertiary period opened.

Let us first consider the fundamental and physical aspect of this revolution, the upheaval of the land. It began about the close of the Jurassic period. Western and Central Europe emerged considerably from the warm Jurassic sea, which lay on it and had converted it into an archipelago. In North-western America also there was an emergence of large areas of land, and the Sierra and Cascade ranges of mountains were formed about the same time. For reasons which will appear later we must note carefully this rise of land at the very beginning of the Cretaceous period.

However, the sea recovered its lost territory, or compensation for it, and the middle of the Cretaceous period witnessed a very considerable extension of the waters over America, Europe, and southern Asia. The thick familiar beds of chalk, which stretch irregularly from Ireland to the Crimea, and from the south of Sweden to the south of France, plainly tell of an overlying sea. As is well known, the chalk consists mainly of the shells or outer frames of minute one-celled creatures (Thalamophores) which float in the ocean, and form a deep ooze at its bottom with their discarded skeletons. What depth this ocean must have been is disputed, and hardly concerns us. It is clear that it must have taken an enormous period for microscopic shells to form the thick masses of chalk which cover so much of southern and eastern England. On the lowest estimates the Cretaceous period, which includes the deposit of other strata besides chalk, lasted about three million years. And as people like to have some idea of the time since these things happened, I may add that, on the lowest estimate (which most geologists would at least double), it is about three million years since the last stretches of the chalk-ocean disappeared from the surface of Europe.

But while our chalk cliffs conjure up a vision of England lying deep—at least twenty or thirty fathoms deep—below a warm ocean, in which gigantic Ammonites and Belemnites and sharks ply their deadly trade, they also remind us of the last phase of the remarkable life of the earth's Middle Ages. In the latter part of the Cretaceous the land rises. The chalk ocean of Europe is gradually reduced to a series of inland seas, separated by masses and ridges of land, and finally to a series of lakes of brackish water. The masses of the Pyrenees and Alps begin to rise; though it will not be until a much later date that they reach anything like their present elevation. In America the change is even greater. A vast ridge rises along the whole western front of the continent, lifting and draining it, from Alaska to Cape Horn. It is the beginning of the Rocky Mountains and the Andes. Even during the Cretaceous period there had been rich forests of Mesozoic vegetation covering about a hundred thousand square miles in the Rocky Mountains region. Europe and America now begin to show their modern contours.

It is important to notice that this great uprise of the land and the series of disturbances it entails differ from those which we summed up in the phrase Permian Revolution. The differences may help us to understand some of the changes in the living population. The chief difference is that the disturbances are more local, and not nearly simultaneous. There is a considerable emergence of land at the end of the Jurassic, then a fresh expansion of the sea, then a great rise of mountains at the end of the Cretaceous, and so on. We shall find our great mountain-masses (the Pyrenees, Alps, Himalaya, etc.) rising at intervals throughout the whole of the Tertiary Era. However, it suffices for the moment to observe that in the latter part of the Mesozoic and early part of the Tertiary there were considerable upheavals of the land in various regions, and that the Mesozoic Era closed with a very much larger proportion of dry land, and a much higher relief of the land, than there had been during the Jurassic period. The series of disturbances was, says Professor Chamberlin, "greater than any that had occurred since the close of the Palaeozoic."

From the previous effect of the Permian upheaval, and from the fact that the living population is now similarly annihilated or reduced, we should at once expect to find a fresh change in the climate of the earth. Here, however, our procedure is not so easy. In the Permian age we had solid proof in the shape of vast glaciated regions. It is claimed by continental geologists that certain early Tertiary beds in Bavaria actually prove a similar, but smaller, glaciation in Europe, but this is disputed. Other beds may yet be found, but we saw that there was not a general upheaval, as there had been in the Permian, and it is quite possible that there were few or no ice-fields. We do not, in fact, know the causes of the Permian icefields. We are thrown upon the plant and animal remains, and seem to be in some danger of inferring a cold climate from the organic remains, and then explaining the new types of organisms by the cold climate. This, of course, we shall not do. The difficulty is made greater by the extreme disinclination of many recent geologists, and some recent botanists who have too easily followed the geologists, to admit a plain climatic interpretation of the facts. Let us first see what the facts are.

In the latter part of the Jurassic we find three different zones of Ammonites: one in the latitude of the Mediterranean, one in the latitude of Central Europe, and one further north. Most geologists conclude that these differences indicate zones of climate (not hitherto indicated), but it cannot be proved, and we may leave the matter open. At the same time the warm-loving corals disappear from Europe, with occasional advances. It is said that they are driven out by the disturbance of the waters, and, although this would hardly explain why they did not spread again in the tranquil chalk-ocean, we may again leave the point open.

In the early part of the Cretaceous, however, the Angiosperms (flowering plants) suddenly break into the chronicle of the earth, and spread with great rapidity. They appear abruptly in the east of the North American continent, in the region of Virginia and Maryland. They are small in stature and primitive in structure. Some are of generalised forms that are now unknown; some have leaves approaching those of the oak, willow, elm, maple, and walnut; some may be definitely described as fig, sassafras, aralia, myrica, etc. Eastern America, it may be recalled, is much higher than western until the close of the Cretaceous period. The Angiosperms do not spread much westward; they appear next in Greenland, and, before the middle of the Cretaceous, in Portugal. They have travelled over the North Atlantic continent, or what remains of it. The process seems very rapid as we write it, but it must be remembered that the first half of the Cretaceous period means a million or a million and a half years.

The cycads, and even the conifers, shrink before the higher type of tree. The landscape, in Europe and America, begins to wear a modern aspect. Long before the end of the Cretaceous most of the modern genera of Angiosperm trees have developed. To the fig and sassafras are now added the birch, beech, oak, poplar, walnut, willow, ivy, mulberry, holly, laurel, myrtle, maple, oleander, magnolia, plane, bread-fruit, and sweet-gum. Most of the American trees of to-day are known. The sequoias (the giant Californian trees) still represent the conifers in great abundance, with the eucalyptus and other plants that are now found only much further south. The ginkgoes struggle on for a time. The cycads dwindle enormously. Of 700 specimens in one early Cretaceous deposit only 96 are Angiosperms; of 460 species in a later deposit about 400 are Angiosperms. They oust the cycads in Europe and America, as the cycads and conifers had ousted the Cryptogams. The change in the face of the earth would be remarkable. Instead of the groves of palm-like cycads, with their large and flower-like fructifications, above which the pines and firs and cypresses reared their sombre forms, there were now forests of delicate-leaved maples, beeches, and oaks, bearing nutritious fruit for the coming race of animals. Grasses also and palms begin in the Cretaceous; though the grasses would at first be coarse and isolated tufts. Even flowers, of the lily family (apparently), are still detected in the crushed and petrified remains.

We will give some consideration later to the evolution of the Angiosperms. For the moment it is chiefly important to notice a feature of them to which the botanist pays less attention. In his technical view the Angiosperm is distinguished by the structure of its reproductive apparatus, its flowers, and some recent botanists wonder whether the key to this expansion of the flowering plants may not be found in a development of the insect world and of its relation to vegetation. In point of fact, we have no geological indication of any great development of the insects until the Tertiary Era, when we shall find them deploying into a vast army and producing their highest types. In any case, such a view leaves wholly unexplained the feature of the Angiosperms which chiefly concerns us. This is that most of them shed the whole of their leaves periodically, as the winter approaches. No such trees had yet been known on the earth. All trees hitherto had been evergreen, and we need a specific and adequate explanation why the earth is now covered, in the northern region, with forests of trees which show naked boughs and branches during a part of the year.

The majority of palaeontologists conclude at once, and quite confidently, from this rise and spread of the deciduous trees, that a winter season has at length set in on the earth, and that this new type of vegetation appears in response to an appreciable lowering of the climate. The facts, however, are somewhat complex, and we must proceed with caution. It would seem that any general lowering of the temperature of the earth ought to betray itself first in Greenland, but the flora of Greenland remains far "warmer," so to say, than the flora of Central Europe is to-day. Even toward the close of the Cretaceous its plants are much the same as those of America or of Central Europe. Its fossil remains of that time include forty species of ferns, as well as cycads, ginkgoes, figs, bamboos, and magnolias. Sir A. Geikie ventures to say that it must then have enjoyed a climate like that of the Cape or of Australia to-day. Professor Chamberlin finds its flora like that of "warm temperate" regions, and says that plants which then flourished in latitude 72 degrees are not now found above latitude 30 degrees.

There are, however, various reasons to believe that it is unsafe to draw deductions from the climate of Greenland. There is, it is true, some exaggeration in the statement that its climate was equivalent to that of Central Europe. The palms which flourished in Central Europe did not reach Greenland, and there are differences in the northern Molluscs and Echinoderms which—like the absence of corals above the north of England—point to a diversity of temperature. But we have no right to expect that there would be the same difference in temperature between Greenland and Central Europe as we find to-day. If the warm current which is now diverted to Europe across the Atlantic— the Gulf Stream—had then continued up the coast of America, and flowed along the coast of the land that united America and Europe, the climatic conditions would be very different from what they are. There is a more substantial reason. We saw that during the Mesozoic the Arctic continent was very largely submerged, and, while Europe and America rise again at the end of the Cretaceous, we find no rise of the land further north. A difference of elevation would, in such a world, make a great difference in temperature and moisture.

Let us examine the animal record, however, before we come to any conclusion. The chronicle of the later Cretaceous is a story of devastation. The reduction of the cyeads is insignificant beside the reduction or annihilation of the great animals of the Mesozoic world. The skeletons of the Deinosaurs become fewer and fewer as we ascend the upper Cretaceous strata. In the uppermost layer (Laramie) we find traces of a last curious expansion—the group of horned reptiles, of the Triceratops type, which we described as the last of the great reptiles. The Ichthyosaurs and Plesiosaurs vanish from the waters. The "sea-serpents" (Mososaurs) pass away without a survivor. The flying dragons, large and small, become entirely extinct. Only crocodiles, lizards, turtle, and snakes cross the threshold of the Tertiary Era. In one single region of America (Puerco beds) some of the great reptiles seem to be making a last stand against the advancing enemy in the dawn of the Tertiary Era, but the exact date of the beds is disputed, and in any case their fight is soon over. Something has slain the most formidable race that the earth had yet known, in spite of its marvellous adaptation to different environments in its innumerable branches.

We turn to the seas, and find an equal carnage among some of its most advanced inhabitants. The great cuttlefish-like Belemnites and the whole race of the Ammonites, large and small, are banished from the earth. The fall of the Ammonites is particularly interesting, and has inspired much more or less fantastic speculation. The shells begin to assume such strange forms that observers speak occasionally of the "convulsions" or "death-contortions" of the expiring race. Some of the coiled shells take on a spiral form, like that of a snail's shell. Some uncoil the shell, and seem to be returning toward the primitive type. A rich eccentricity of frills and ornamentation is found more or less throughout the whole race. But every device—if we may so regard these changes—is useless, and the devastating agency of the Cretaceous, whatever it was, removes the Ammonites and Belemnites from the scene. The Mollusc world, like the world of plants and of reptiles, approaches its modern aspect.

In the fish world, too, there is an effective selection in the course of the Cretaceous. All the fishes of modern times, except the large family of the sharks, rays, skates, and dog-fishes (Elasmobranchs), the sturgeon and chimaera, the mud-fishes, and a very few other types, are Teleosts, or bony-framed fishes—the others having cartilaginous frames. None of the Teleosts had appeared until the end of the Jurassic. They now, like the flowering plants on land, not only herald the new age, but rapidly oust the other fishes, except the unconquerable shark. They gradually approach the familiar types of Teleosts, so that we may say that before the end of the Cretaceous the waters swarmed with primitive and patriarchal cod, salmon, herring, perch, pike, bream, eels, and other fishes. Some of them grew to an enormous size. The Portheus, an American pike, seems to have been about eight feet long; and the activity of an eight-foot pike may be left to the angler's imagination. All, however, are, as evolution demands, of a generalised and unfamiliar type: the material out of which our fishes will be evolved.

Of the insects we have very little trace in the Cretaceous. We shall find them developing with great richness in the following period, but, imperfect as the record is, we may venture to say that they were checked in the Cretaceous. There were good conditions for preserving them, but few are preserved. And of the other groups of invertebrates we need only say that they show a steady advance toward modern types. The sea-lily fills the rocks no longer; the sea-urchin is very abundant. The Molluscs gain on the more lowly organised Brachiopods.

To complete the picture we must add that higher types probably arose in the later Cretaceous which do not appear in the records. This is particularly true of the birds and mammals. We find them spreading so early in the Tertiary that we must put back the beginning of the expansion to the Cretaceous. As yet, however, the only mammal remains we find are such jaws and teeth of primitive mammals as we have already described. The birds we described (after the Archaeopteryx) also belong to the Cretaceous, and they form another of the doomed races. Probably the modern birds were already developing among the new vegetation on the higher ground.

These are the facts of Cretaceous life, as far as the record has yielded them, and it remains for us to understand them. Clearly there has been a great selective process analogous to, if not equal to, the winnowing process at the end of the Palaeozoic. As there has been a similar, if less considerable, upheaval of the land, we are at once tempted to think that the great selective agency was a lowering of the temperature. When we further find that the most important change in the animal world is the destruction of the cold-blooded reptiles, which have no concern for the young, and the luxuriant spread of the warm-blooded animals, which do care for their young, the idea is greatly confirmed. When we add that the powerful Molluscs which are slain, while the humbler Molluscs survive, are those which—to judge from the nautilus and octopus—love warm seas, the impression is further confirmed. And when we finally reflect that the most distinctive phenomenon of the period is the rapid spread of deciduous trees, it would seem that there is only one possible interpretation of the Cretaceous Revolution.

This interpretation—that cold was the selecting agency—is a familiar idea in geological literature, but, as I said, there are recent writers who profess reserve in regard to it, and it is proper to glance at, or at least look for, the alternatives.

Before doing so let us be quite clear that here we have nothing to do with theories of the origin of the earth. The Permian cold—which, however, is universally admitted—is more or less entangled in that controversy; the Cretaceous cold has no connection with it. Whatever excess of carbon-dioxide there may have been in the early atmosphere was cleared by the Coal-forests. We must set aside all these theories in explaining the present facts.

It is also useful to note that the fact that there have been great changes in the climate of the earth in past time is beyond dispute. There is no denying the fact that the climate of the earth was warm from the Arctic to the Antarctic in the Devonian and Carboniferous periods: that it fell considerably in the Permian: that it again became at least "warm temperate" (Chamberlin) from the Arctic to the Antarctic in the Jurassic, and again in the Eocene: that some millions of square miles of Europe and North America were covered with ice and snow in the Pleistocene, so that the reindeer wandered where palms had previously flourished and the vine flourishes to-day; and that the pronounced zones of climate which we find today have no counterpart in any earlier age. In view of these great and admitted fluctuations of the earth's temperature one does not see any reason for hesitating to admit a fall of temperature in the Cretaceous, if the facts point to it.

On the other hand, the alternative suggestions are not very convincing. We have noticed one of these suggestions in connection with the origin of the Angiosperms. It hints that this may be related to developments of the insect world. Most probably the development of the characteristic flowers of the Angiosperms is connected with an increasing relation to insects, but what we want to understand especially is the deciduous character of their leaves. Many of the Angiosperms are evergreen, so that it cannot be said that the one change entailed the other. In fact, a careful study of the leaves preserved in the rocks seems to show the deciduous Angiosperms gaining on the evergreens at the end of the Cretaceous. The most natural, it not the only, interpretation of this is that the temperature is falling. Deciduous trees shed their leaves so as to check their transpiration when a season comes on in which they cannot absorb the normal amount of moisture. This may occur either at the on-coming of a hot, dry season or of a cold season (in which the roots absorb less). Everything suggests that the deciduous tree evolved to meet an increase of cold, not of heat.

Another suggestion is that animals and plants were not "climatically differentiated" until the Cretaceous period; that is to say, that they were adapted to all climates before that time, and then began to be sensitive to differences of climate, and live in different latitudes. But how and why they should suddenly become differentiated in this way is so mysterious that one prefers to think that, as the animal remains also suggest, there were no appreciable zones of climate until the Cretaceous. The magnolia, for instance, flourished in Greenland in the early Tertiary, and has to live very far south of it to-day. It is much simpler to assume that Greenland changed—as a vast amount of evidence indicates—than that the magnolia changed.

Finally, to explain the disappearance of the Mesozoic reptiles without a fall in temperature, it is suggested that they were exterminated by the advancing mammals. It is assumed that the spreading world of the Angiospermous plants somewhere met the spread of the advancing mammals, and opened out a rich new granary to them. This led to so powerful a development of the mammals that they succeeded in overthrowing the reptiles.

There are several serious difficulties in the way of this theory. The first and most decisive is that the great reptiles have practically disappeared before the mammals come on the scene. Only in one series of beds (Puerco) in America, representing an early period of the Tertiary Era, do we find any association of their remains; and even there it is not clear that they were contemporary. Over the earth generally the geological record shows the great reptiles dying from some invisible scourge long before any mammal capable of doing them any harm appears; even if we suppose that the mammal mainly attacked the eggs and the young. We may very well believe that more powerful mammals than the primitive Mesozoic specimens were already developed in some part of the earth—say, Africa—and that the rise of the land gave them a bridge across the Mediterranean to Europe. Probably this happened; but the important point is that the reptiles were already almost extinct. The difficulty is even greater when we reflect that it is precisely the most powerful reptiles (Deinosaurs) and least accessible reptiles (Pterosaurs, Ichthyosaurs, etc.) which disappear, while the smaller land and water reptiles survive and retreat southward—where the mammals are just as numerous. That assuredly is not the effect of an invasion of carnivores, even if we could overlook the absence of such carnivores from the record until after the extinction of the reptiles in most places.

I have entered somewhat fully into this point, partly because of its great interest, but partly lest it be thought that I am merely reproducing a tradition of geological literature without giving due attention to the criticisms of recent writers. The plain and common interpretation of the Cretaceous revolution—that a fall in temperature was its chief devastating agency—is the only one that brings harmony into all the facts. The one comprehensive enemy of that vast reptile population was cold. It was fatal to the adult because he had a three-chambered heart and no warm coat; it was fatal to the Mesozoic vegetation on which, directly or indirectly, he fed; it was fatal to his eggs and young because the mother did not brood over the one or care for the other. It was fatal to the Pterosaurs, even if they were warm-blooded, because they had no warm coats and did not (presumably) hatch their eggs; and it was equally fatal to the viviparous Ichthyosaurs. It is the one common fate that could slay all classes. When we find that the surviving reptiles retreat southward, only lingering in Europe during the renewed warmth of the Eocene and Miocene periods, this interpretation is sufficiently confirmed. And when we recollect that these things coincide with the extinction of the Ammonites and Belemnites, and the driving of their descendants further south, as well as the rise and triumph of deciduous trees, it is difficult to see any ground for hesitating.

But we need not, and must not, imagine a period of cold as severe, prolonged, and general as that of the Permian period. The warmth of the Jurassic period is generally attributed to the low relief of the land, and the very large proportion of water-surface. The effect of this would be to increase the moisture in the atmosphere. Whether this was assisted by any abnormal proportion of carbon-dioxide, as in the Carboniferous, we cannot confidently say. Professor Chamberlin observes that, since the absorbing rock-surface was greatly reduced in the Jurassic, the carbon-dioxide would tend to accumulate in its atmosphere, and help to explain the high temperature. But the great spread of vegetation and the rise of land in the later

236

Jurassic and the Cretaceous would reduce this density of the atmosphere, and help to lower the temperature.

It is clear that the cold would at first be local. In fact, it must be carefully realised that, when we speak of the Jurassic period as a time of uniform warmth, we mean uniform at the same altitude. Everybody knows the effect of rising from the warm, moist sea-level to the top of even a small inland elevation. There would be such cooler regions throughout the Jurassic, and we saw that there were considerable upheavals of land towards its close. To these elevated lands we may look for the development of the Angiosperms, the birds, and the mammals. When the more massive rise of land came at the end of the Cretaceous, the temperature would fall over larger areas, and connecting ridges would be established between one area and another. The Mesozoic plants and animals would succumb to this advancing cold. What precise degree of cold was necessary to kill the reptiles and Cephalopods, yet allow certain of the more delicate flowering plants to live, is yet to be determined. The vast majority of the new plants, with their winter sleep, would thrive in the cooler air, and, occupying the ground of the retreating cycads and ginkgoes would prepare a rich harvest for the coming birds and mammals.

CHAPTER XV. THE TERTIARY ERA

We have already traversed nearly nine-tenths of the story of terrestrial life, without counting the long and obscure Archaean period, and still find ourselves in a strange and unfamiliar earth. With the close of the Chalk period, however, we take a long stride in the direction of the modern world. The Tertiary Era will, in the main, prove a fresh period of genial warmth and fertile low-lying regions. During its course our deciduous trees and grasses will mingle with the palms and pines over the land, our flowers will begin to brighten the landscape, and the forms of our familiar birds and mammals, even the form of man, will be discernible in the crowds of animals. At its close another mighty period of selection will clear the stage for its modern actors.

A curious reflection is prompted in connection with this division of the earth's story into periods of relative prosperity and quiescence, separated by periods of disturbance. There was—on the most modest estimate—a stretch of some fifteen million years between the Cambrian and the Permian upheavals. On the same chronological scale the interval between the Permian and Cretaceous revolutions was only about seven million years, and the Tertiary Era will comprise only about three million years. One wonders if the Fourth (Quaternary) Era in which we live will be similarly shortened. Further, whereas the earth returned after each of the earlier upheavals to what seems to

238

have been its primitive condition of equable and warm climate, it has now entirely departed from that condition, and exhibits very different zones of climate and a succession of seasons in the year. One wonders what the climate of the earth will become long before the expiration of those ten million years which are usually assigned as the minimum period during which the globe will remain habitable.

It is premature to glance at the future, when we are still some millions of years from the present, but it will be useful to look more closely at the facts which inspire this reflection. From what we have seen, and shall further see, it is clear that, in spite of all the recent controversy about climate among our geologists, there has undeniably been a progressive refrigeration of the globe. Every geologist, indeed, admits "oscillations of climate," as Professor Chamberlin puts it. But amidst all these oscillations we trace a steady lowering of the temperature. Unless we put a strained and somewhat arbitrary interpretation on the facts of the geological record, earlier ages knew nothing of our division of the year into pronounced seasons and of the globe into very different climatic zones. It might plausibly be suggested that we are still living in the last days of the Ice-Age, and that the earth may be slowly returning to a warmer condition. Shackleton, it might be observed, found that there has been a considerable shrinkage of the south polar ice within the period of exploration. But we shall find that a difference of climate, as compared with earlier ages, was already evident in the middle of the Tertiary Era, and it is far more noticeable to-day.

We do not know the causes of this climatic evolution—the point will be considered more closely in connection with the last Ice-Age—but we see that it throws a flood of light on the evolution of organisms. It is one of the chief incarnations of natural selection. Changes in the distribution of land and water and in the nature of the land-surface, the coming of powerful carnivores, and other agencies which we have seen, have had their share in the onward impulsion of life, but the most drastic agency seems to have been the supervention of cold. The higher types of both animals and plants appear plainly in response to a lowering of temperature. This is the chief advantage of studying the story of evolution in strict connection with the geological record. We shall find that the record will continue to throw light on our path to the end, but, as we are now about to approach the most important era of evolution, and as we have now seen so much of the concrete story of evolution, it will be interesting to examine briefly some other ways of conceiving that story.

We need not return to the consideration of the leading schools of evolution, as described in a former chapter. Nothing that we have seen will enable us to choose between the Lamarckian and the Weismannist hypothesis; and I doubt if anything we are yet to see will prove more decisive. The dispute is somewhat academic, and not vital to a conception of evolution. We shall, for instance, presently follow the evolution of the horse, and see four of its toes shrink and disappear, while the fifth toe is enormously strengthened. In the facts themselves there is nothing whatever to decide whether this evolution took place on the lines suggested by Weismann, or on the lines suggested by Lamarck and accepted by Darwin. It will be enough for us merely to establish the fact that the one-toed horse is an evolved descendant of a primitive five-toed mammal, through the adaptation of its foot to running on firm ground, its teeth and neck to feeding on grasses, and so on.

On the other hand, the facts we have already seen seem to justify the attitude of compromise I adopted in regard to the Mutationist theory. It would be an advantage in many ways if we could believe that new species arose by sudden and large variations (mutations) of the young from the parental type. In the case of many organs and habits it is extremely difficult to see how a gradual development, by a slow accentuation of small variations, is possible. When we further find that experimenters on living species can bring about such mutations, and when we reflect that there must have been acute disturbances in the surroundings of animals and plants sometimes, we are disposed to think that many a new species may have arisen in this way. On the other hand, while the palaeontological record can never prove that a species arose by mutations, it does sometimes show that species arise by very gradual modification. The Chalk period, which we have just traversed, affords a very clear instance. One of our chief investigators of the English Chalk, Dr. Rowe, paid particular attention to the sea-urchins it contains, as they serve well to identify different levels of chalk. He discovered, not merely that they vary from level to level, but that in at least one genus (Micraster) he could trace the organism very gradually passing from one species to another, without any leap or abruptness. It is certainly significant that we find such cases as this precisely where the conditions of preservation are exceptionally good. We must conclude that species arise, probably, both by mutations and small variations, and that it is impossible to say which class of species has been the more numerous.

There remain one or two conceptions of evolution which we have not hitherto noticed, as it was advisable to see the facts first. One of these is the view—chiefly represented in this country by Professor Henslow—that natural selection has had no part in the creation of species; that the only two factors are the environment and the organism which responds to its changes. This is true enough in the sense that, as we saw, natural selection is not an action of nature on the "fit," but on the unfit or less fit. But this does not in the least lessen the importance of natural selection. If there were not in nature this body of destructive agencies, to which we apply the name natural selection, there would be little—we cannot say no—evolution. But the rising carnivores, the falls of temperature, etc., that we have studied, have had so real, if indirect, an influence on the development of life that we need not dwell on this.

Another school, or several schools, while admitting the action of natural selection, maintain that earlier evolutionists have made nature much too red in tooth and claw. Dr. Russel Wallace from one motive, and Prince Krapotkin from another, have insisted that the triumphs of war have been exaggerated, and the triumphs of peace, or of social co-operation, far too little appreciated. It will be found that such writers usually base their theory on life as we find it in nature to-day, where the social principle is highly developed in many groups of animals. This is most misleading, since social co-operation among animals, as an instrument of progress, is (geologically speaking) quite a recent phenomenon. Nearly every group of animals in which it is found belongs, to put it moderately, to the last tenth of the story of life, and in some of the chief instances the animals have only gradually developed social life. [*] The first nine-tenths of the chronicle of evolution contain no indication of social life, except—curiously enough—in such groups as the Sponges, Corals, and Bryozoa, which are amongst the least progressive in nature. We have seen plainly that during the overwhelmingly greater part of the story of life the predominant agencies of evolution were struggle against adverse conditions and devouring carnivores; and we shall find them the predominant agencies throughout the Tertiary Era.

* Thus the social nature of man is sometimes quoted as one
     of the chief causes of his development. It is true that it
     has much to do with his later development, but we shall see
     that the statement that man was from the start a social
     being is not at all warranted by the facts. On the other
     hand, it may be pointed out that the ants and termites had
     appeared in the Mesozoic. We shall see some evidence that
     the remarkable division of labour which now characterises

Yet we must protest against the exaggerated estimate of the conscious pain which so many read into these millions of years of struggle. Probably there was no consciousness at all during the greater part of the time. The wriggling of the worm on which you have accidentally trodden is no proof whatever that you have caused conscious pain. The nervous system of an animal has been so evolved as to respond with great disturbance of its tissue to any dangerous or injurious assault. It is the selection of a certain means of self-preservation. But at what level of life the animal becomes conscious of this disturbance, and "feels pain," it is very difficult to determine. The subject is too vast to be opened here. In a special investigation of it. [*] I concluded that there is no proof of the presence of any degree of consciousness in the invertebrate world even in the higher insects; that there is probably only a dull, blurred, imperfect consciousness below the level of the higher mammals and birds; and that even the consciousness of an ape is something very different from what educated Europeans, on the ground of their own experience, call consciousness. It is too often forgotten that pain is in proportion to consciousness. We must beware of such fallacies as transferring our experience of pain to a Mesozoic reptile, with an ounce or two of cerebrum to twenty tons of muscle and bone.

* *"The Evolution of Mind" (Black), 1911.*

One other view of evolution, which we find in some recent and reputable works (such as Professor Geddes and Thomson's "Evolution," 1911), calls for consideration. In the ordinary Darwinian view the variations of the young from their parents are indefinite, and spread in all directions. They may continue to occur for ages without any of them proving an advantage to their possessors. Then the environment may change, and a certain variation may prove an advantage, and be continuously and increasingly selected. Thus these indefinite variations may be so controlled by the environment during millions of years that the fish at last becomes an elephant or a man. The alternative view, urged by a few writers, is that the variations were "definitely directed." The phrase seems merely to complicate the story of evolution with a fresh and superfluous mystery. The nature and precise action of this "definite direction" within the organism are quite unintelligible, and the facts seem explainable just as well—or not less imperfectly—without as with this mystic agency. Radiolaria, Sponges, Corals, Sharks, Mudfishes, Duckbills, etc., do not change (except within the limits of their family) during millions of years, because they keep to an environment to which they are fitted. On the other hand, certain fishes, reptiles, etc., remain in a changing environment, and they must change with it. The process has its obscurities, but we make them darker, it seems to me, with these semi-metaphysical phrases.

It has seemed advisable to take this further glance at the general principles and current theories of evolution before we extend our own procedure into the Tertiary Era. The highest types of animals and plants are now about to appear on the stage of the earth; the theatre

245

itself is about to take on a modern complexion. The Middle Ages are over; the new age is breaking upon the planet. We will, as before, first survey the Tertiary Era as a whole, with the momentous changes it introduces, and then examine, in separate chapters, the more important phases of its life.

It opens, like the preceding and the following era, with "the area of land large and its relief pronounced." This is the outcome of the Cretaceous revolution. Southern Europe and Southern Asia have risen, and shaken the last masses of the Chalk ocean from their faces; the whole western fringe of America has similarly emerged from the sea that had flooded it. In many parts, as in England (at that time a part of the Continent), there is so great a gap between the latest Cretaceous and the earliest Tertiary strata that these newly elevated lands must evidently have stood out of the waters for a prolonged period. On their cooler plains the tragedy of the extinction of the great reptiles comes to an end. The cyeads and ginkgoes have shrunk into thin survivors of the luxuriant Mesozoic groves. The oak and beech and other deciduous trees spread slowly over the successive lands, amid the glare and thunder of the numerous volcanoes which the disturbance of the crust has brought into play. New forms of birds fly from tree to tree, or linger by the waters; and strange patriarchal types of mammals begin to move among the bones of the stricken reptiles.

But the seas and the rains and rivers are acting with renewed vigour on the elevated lands, and the Eocene period closes in a fresh age of levelling. Let us put the work of a million years or so in a sentence. The southern sea, which has been confined almost to the limits of our Mediterranean by the Cretaceous upheaval, gradually enlarges once more. It floods the north-west of Africa almost as far as the equator; it covers most of Italy, Turkey, Austria, and Southern Russia; it spreads over Asia Minor, Persia, and Southern Asia, until it joins the Pacific; and it sends a long arm across the Franco-British region, and up the great valley which is now the German Ocean.

From earlier chapters we now expect to find a warmer climate, and the record gives abundant proof of it. To this period belongs the "London Clay," in whose thick and—to the unskilled eye—insignificant bed the geologist reads the remarkable story of what London was two or three million years ago. It tells us that a sea, some 500 or 600 feet deep, then lay over that part of England, and fragments of the life of the period are preserved in its deposit. The sea lay at the mouth of a sub-tropical river on whose banks grew palms, figs, ginkgoes, eucalyptuses, almonds, and magnolias, with the more familiar oaks and pines and laurels. Sword-fishes and monstrous sharks lived in the sea. Large turtles and crocodiles and enormous "sea-serpents" lingered in this last spell of warmth that Central Europe would experience. A primitive whale appeared in the seas, and strange large tapir-like mammals—remote ancestors of our horses and more familiar beasts—wandered heavily on the land. Gigantic primitive birds, sometimes ten feet high, waded by the shore. Deposits of the period at Bournemouth and in the Isle of Wight tell the same story of a land that bore figs, vines, palms, araucarias, and aralias, and waters that sheltered turtles and crocodiles. The Parisian region presented the same features.

In fact, one of the most characteristic traces of the southern sea which then stretched from England to Africa in the south and India in the east indicates a warm climate. It will be remembered that the Cretaceous ocean over Southern Europe had swarmed with the animalcules whose dead skeletons largely compose our chalk-beds. In the new southern ocean another branch of these Thalamophores, the Nummulites, spreads with such portentous abundance that its shells—sometimes alone, generally with other material—make beds of solid limestone several thousand feet in thickness. The pyramids are built of this nummulitic limestone. The one-celled animal in its shell is, however, no longer a microscopic grain. It sometimes forms wonderful shells, an inch or more in diameter, in which as many as a thousand chambers succeed each other, in spiral order, from the centre. The beds containing it are found from the Pyrenees to Japan.

That this vast warm ocean, stretching southward over a large part of what is now the Sahara, should give a semitropical aspect even to Central Europe and Asia is not surprising. But this genial climate was still very general over the earth. Evergreens which now need the warmth of Italy or the Riviera then flourished in Lapland and Spitzbergen. The flora of Greenland—a flora that includes magnolias, figs, and bamboos—shows us that its temperature in the Eocene period must have been about 30 degrees higher than it is to-day. [*] The temperature of the cool Tyrol of modern Europe is calculated to have then been between 74 and 81 degrees F. Palms, cactuses, aloes, gum-trees, cinnamon trees, etc., flourished in the latitude of Northern France. The forests that covered parts of Switzerland which are now buried in snow during a great part of the year were like the forests one finds in parts of India and Australia to-day. The climate of North America, and of the land which still connected it with Europe, was correspondingly genial.

```
    * The great authority on Arctic geology, Heer, who
makes
    this calculation, puts this flora in the Miocene.
It is now
    usually considered that these warmer plants belong
to the
    earlier part of the Tertiary era.
```

This indulgent period (the Oligocene, or later part of the Eocene), scattering a rich and nutritious vegetation with great profusion over the land, led to a notable expansion of animal life. Insects, birds, and mammals spread into vast and varied groups in every land. Had any of the great Mesozoic reptiles survived, the warmer age might have enabled them to dispute the sovereignty of the advancing mammals. But nothing more formidable than the turtle, the snake, and the crocodile (confined to the waters) had crossed the threshold of the Tertiary Era, and the mammals and birds had the full advantage of the new golden age. The fruits of the new trees, the grasses which now covered the plains, and the insects which multiplied with the flowers afforded a magnificent diet. The herbivorous mammals became a populous world, branching into numerous different types according to their different environments. The horse, the elephant, the camel, the pig, the deer, the rhinoceros gradually emerge out of the chaos of evolving forms. Behind them, hastening the course of their evolution, improving their speed, arms, and armour, is the inevitable carnivore. He, too, in the abundance of food, grows into a vast population, and branches out toward familiar types. We will devote a chapter presently to this remarkable phase of the story of evolution.

But the golden age closes, as all golden ages had done before it, and for the same reason. The land begins to rise, and cast the warm shallow seas from its face. The expansion of life has been more rapid and remarkable than it had ever been before, in corresponding periods of abundant food and easy conditions; the contraction comes more quickly than it had ever done before. Mountain masses begin to rise in nearly all parts of the world. The advance is slow and not continuous, but as time goes on the Atlas, Alps, Pyrenees, Apennines, Caucasus, Himalaya, Rocky Mountains, and Andes rise higher and higher. When the geologist looks to-day for the floor of the Eocene ocean, which he recognises by the shells of the Nummulites, he finds it 10,000 feet above the sea-level in the Alps, 16,000 feet above the sea-level in the Himalaya, and 20,000 feet above the sea-level in Thibet. One need not ask why the regions of London and Paris fostered palms and magnolias and turtles in Tertiary times, and shudder in their dreary winter to-day.

The Tertiary Era is divided by geologists into four periods: the Eocene, Oligocene, Miocene, and

Pliocene. "Cene" is our barbaric way of expressing the Greek word for "new," and the classification is meant to mark the increase of new (or modern and actual) types of life in the course of the Tertiary Era. Many geologists, however, distrust the classification, and are disposed to divide the Tertiary into two periods. From our point of view, at least, it is advisable to do this. The first and longer half of the Tertiary is the period in which the temperature rises until Central Europe enjoys the climate of South Africa; the second half is the period in which the land gradually rises, and the temperature falls, until glaciers and sheets of ice cover regions where the palm and fig had flourished.

The rise of the land had begun in the first half of the Tertiary, but had been suspended. The Pyrenees and Apennines had begun to rise at the end of the Eocene, straining the crust until it spluttered with volcanoes, casting the nummulitic sea off large areas of Southern Europe. The Nummulites become smaller and less abundant. There is also some upheaval in North America, and a bridge of land begins to connect the north and south, and permit an effective mingling of their populations. But the advance is, as I said, suspended, and the Oligocene period maintains the golden age. With the Miocene period the land resumes its rise. A chill is felt along the American coast, showing a fall in the temperature of the Atlantic. In Europe there is a similar chill, and a more obvious reason for it. There is an ascending movement of the whole series of mountains from Morocco and the Pyrenees, through the Alps, the Caucasus, and the Carpathians, to India and China. Large lakes still lie over Western Europe, but nearly the whole of it emerges from the ocean. The Mediterranean still sends an arm up France, and with another arm encircles the Alpine mass; but the upheaval continues, and the great nummulitic sea is reduced to a series of extensive lakes, cut off both from the Atlantic and Pacific. The climate of Southern Europe is probably still as genial as that of the Canaries to-day. Palms still linger in the landscape in reduced numbers.

The last part of the Tertiary, the Pliocene, opens with a slight return of the sea. The upheaval is once more suspended, and the waters are eating into the land. There is some foundering of land at the south-western tip of Europe; the "Straits of Gibraltar" begin to connect the Mediterranean with the Atlantic, and the Balearic Islands, Corsica, and Sardinia remain as the mountain summits of a submerged land. Then the upheaval is resumed, in nearly every part of the earth.

Nearly every great mountain chain that the geologist has studied shared in this remarkable movement at the end of the Tertiary Era. The Pyrenees, Alps, Himalaya, etc., made their last ascent, and attained their present elevation. And as the land rose, the aspect of Europe and America slowly altered. The palms, figs, bamboos, and magnolias disappeared; the turtles, crocodiles, flamingoes, and hippopotamuses retreated toward the equator. The snow began to gather thick on the rising heights; then the glaciers began to glitter on their flanks. As the cold increased, the rivers of ice which flowed down the hills of Switzerland, Spain, Scotland, or Scandinavia advanced farther and farther over the plains. The regions of green vegetation shrank before the oncoming ice, the animals retreated south, or developed Arctic features. Europe and America were ushering in the great Ice-Age, which was to bury five or six million square miles of their territory under a thick mantle of ice.

Such is the general outline of the story of the Tertiary Era. We approach the study of its types of life and their remarkable development more intelligently when we have first given careful attention to this extraordinary series of physical changes. Short as the Era is, compared with its predecessors, it is even more eventful and stimulating than they, and closes with what Professor Chamberlin calls "the greatest deformative movements in post-Cambrian history." In the main it has, from the evolutionary point of view, the same significant character as the two preceding eras. Its middle portion is an age of expansion, indulgence, exuberance, in which myriads of varied forms are thrown upon the scene, its later part is an age of contraction, of annihilation, of drastic test, in which the more effectively organised will be chosen from the myriads of types. Once more nature has engendered a vast brood, and is about to select some of her offspring to people the modern world. Among the types selected will be Man.

CHAPTER XVI. THE FLOWER AND THE INSECT

AS we approach the last part of the geological record we must neglect the lower types of life, which have hitherto occupied so much of our attention, so that we may inquire more fully into the origin and fortunes of the higher forms which now fill the stage. It may be noted, in general terms, that they shared the opulence of the mid-Tertiary period, produced some gigantic specimens of their respective families, and evolved into the genera, and often the species, which we find living to-day. A few illustrations will suffice to give some idea of the later development of the lower invertebrates and vertebrates.

Monstrous oysters bear witness to the prosperity of that ancient and interesting family of the Molluscs. In some species the shells were commonly ten inches long; the double shell of one of these Tertiary bivalves has been found which measured thirteen inches in length, eight in width, and six in thickness. In the higher branch of the Mollusc world the naked Cephalopods (cuttle-fish, etc.) predominate over the nautiloids—the shrunken survivors of the great coiled-shell race. Among the sharks, the modern Squalodonts entirely displace the older types, and grow to an enormous size. Some of the teeth we find in Tertiary deposits are more than six inches long and six inches broad at the base. This is three times the size of the teeth of the largest living shark, and it is therefore believed that the extinct possessor of these formidable teeth (Carcharodon megalodon) must have been much more than fifty, and was possibly a hundred, feet in length. He flourished in the waters of both Europe and America during the halcyon days of the Tertiary Era. Among the bony fishes, all our modern and familiar types appear.

The amphibia and reptiles also pass into their modern types, after a period of generous expansion. Primitive frogs and toads make their first appearance in the Tertiary, and the remains are found in European beds of four-foot-long salamanders. More than fifty species of Tertiary turtles are known, and many of them were of enormous size. One carapace that has been found in a Tertiary bed measures twelve feet in length, eight feet in width, and seven feet in height to the top of the back. The living turtle must have been nearly twenty feet long. Marine reptiles, of a snake-like structure, ran to fifteen feet in length. Crocodiles and alligators swarmed in the rivers of Europe until the chilly Pliocene bade them depart to Africa.

In a word, it was the seven years of plenty for the whole living world, and the expansive development gave birth to the modern types, which were to be selected from the crowd in the subsequent seven years of famine. We must be content to follow the evolution of the higher types of organisms. I will therefore first describe the advance of the Tertiary vegetation, the luxuriance of which was the first condition of the great expansion of animal life; then we will glance at the grand army of the insects which followed the development of the flowers, and at the accompanying expansion and ramification of the birds. The long and interesting story of the mammals must be told in a separate chapter, and a further chapter must be devoted to the appearance of the human species.

We saw that the Angiosperms, or flowering plants, appeared at the beginning of the Cretaceous period, and were richly developed before the Tertiary Era opened. We saw also that their precise origin is unknown. They suddenly invade a part of North America where there were conditions for preserving some traces of them, but we have as yet no remains of their early forms or clue to their place of development. We may conjecture that their ancestors had been living in some elevated inland region during the warmth of the Jurassic period.

As it is now known that many of the cycad-like Mesozoic plants bore flowers—as the modern botanist scarcely hesitates to call them—the gap between the Gymnosperms and Angiosperms is very much lessened. There are, however, structural differences which forbid us to regard any of these flowering cycads, which we have yet found, as the ancestors of the Angiosperms. The most reasonable view seems to be that a small and local branch of these primitive flowering plants was evolved, like the rest, in the stress of the Permian-Triassic cold; that, instead of descending to the warm moist levels with the rest at the end of the Triassic, and developing the definite characters of the cycad, it remained on the higher and cooler land; and that the rise of land at the end of the Jurassic period stimulated the development of its Angiosperm features, enlarged the area in which it was especially fitted to thrive, and so permitted it to spread and suddenly break into the geological record as a fully developed Angiosperm.

As the cycads shrank in the Cretaceous period, the Angiosperms deployed with great rapidity, and, spreading at various levels and in different kinds of soils and climates, branched into hundreds of different types. We saw that the oak, beech, elm, maple, palm, grass, etc., were well developed before the end of the Cretaceous period. The botanist divides the Angiosperms into two leading groups, the Monocotyledons (palms, grasses, lilies, orchises, irises, etc.) and Dicotyledons (the vast majority), and it is now generally believed that the former were developed from an early and primitive branch of the latter. But it is impossible to retrace the lines of development of the innumerable types of Angiosperms. The geologist has mainly to rely on a few stray leaves that were swept into the lakes and preserved in the mud, and the evidence they afford is far too slender for the construction of genealogical trees. The student of living plants can go a little further in discovering relationships, and, when we find him tracing such apparently remote plants as the apple and the strawberry to a common ancestor with the rose, we foresee interesting possibilities on the botanical side. But the evolution of the Angiosperms is a recent and immature study, and we will be content with a few reflections on the struggle of the various types of trees in the changing conditions of the Tertiary, the development of the grasses, and the evolution of the flower. In other words, we will be content to ask how the modern landscape obtained its general vegetal features.

Broadly speaking, the vegetation of the first part of the Tertiary Era was a mixture of sub-tropical and temperate forms, a confused mass of Ferns, Conifers, Ginkgoales, Monocotyledons, and Dicotyledons. Here is a casual list of plants that then grew in the latitude of London and Paris: the palm, magnolia, myrtle, Banksia, vine, fig, aralea, sequoia, eucalyptus, cinnamon tree, cactus, agave, tulip tree, apple, plum, bamboo, almond, plane, maple, willow, oak, evergreen oak, laurel, beech, cedar, etc. The landscape must have been extraordinarily varied and beautiful and rich. To one botanist it suggests Malaysia, to another India, to another Australia.

It is really the last gathering of the plants, before the great dispersion. Then the cold creeps slowly down from the Arctic regions, and begins to reduce the variety. We can clearly trace its gradual advance. In the Carboniferous and Jurassic the vegetation of the Arctic regions had been the same as that of England; in the Eocene palms can flourish in England, but not further north; in the Pliocene the palms and bamboos and semi-tropical species are driven out of Europe; in the Pleistocene the ice-sheet advances to the valleys of the Thames and the Danube (and proportionately in the United States), every warmth-loving species is annihilated, and our grasses, oaks, beeches, elms, apples, plums, etc., linger on the green southern fringe of the Continent, and in a few uncovered regions, ready to spread north once more as the ice creeps back towards the Alps or the Arctic circle. Thus, in few

258

words, did Europe and North America come to have the vegetation we find in them to-day.

The next broad characteristic of our landscape is the spreading carpet of grass. The interest of the evolution of the grasses will be seen later, when we shall find the evolution of the horse, for instance, following very closely upon it. So striking, indeed, is the connection between the advance of the grasses and the advance of the mammals that Dr. Russel Wallace has recently claimed ("The World of Life," 1910) that there is a clear purposive arrangement in the whole chain of developments which leads to the appearance of the grasses. He says that "the very puzzling facts" of the immense reptilian development in the Mesozoic can only be understood on the supposition that they were evolved "to keep down the coarser vegetation, to supply animal food for the larger Carnivora, and thus give time for higher forms to obtain a secure foothold and a sufficient amount of varied form and structure" (p. 284).

Every insistence on the close connection of the different strands in the web of life is welcome, but Dr. Wallace does not seem to have learned the facts accurately. There is nothing "puzzling" about the Mesozoic reptilian development; the depression of the land, the moist warmth, and the luscious vegetation of the later Triassic and the Jurassic amply explain it. Again, the only carnivores to whom they seem to have supplied food were reptiles of their own race. Nor can the feeding of the herbivorous reptiles be connected with the rise of the Angiosperms. We do not find the flowering plants developing anywhere in those vast regions where the great reptiles abounded; they invade them from some single unknown region, and mingle with the pines and ginkgoes, while the cyeads alone are destroyed.

The grasses, in particular, do not appear until the Cretaceous, and do not show much development until the mid-Tertiary; and their development seems to be chiefly connected with physical conditions. The meandering rivers and broad lakes of the mid-Tertiary would have their fringes of grass and sedge, and, as the lakes dried up in the vicissitudes of climate, large areas of grass would be left on their sites. To these primitive prairies the mammal (not reptile) herbivores would be attracted, with important results. The consequences to the animals we will consider presently. The effect on the grasses may be well understood on the lines so usefully indicated in Dr. Wallace's book. The incessant cropping, age after age, would check the growth of the larger and coarser grasses give opportunity to the smaller and finer, and lead in time to the development of the grassy plains of the modern world. Thus one more familiar feature was added to the landscape in the Tertiary Era.

As this fresh green carpet spread over the formerly naked plains, it began to be enriched with our coloured flowers. There were large flowers, we saw, on some of the Mesozoic cycads, but their sober yellows and greens—to judge from their descendants—would do little to brighten the landscape. It is in the course of the Tertiary Era that the mantle of green begins to be embroidered with the brilliant hues of our flowers.

Grant Allen put forward in 1882 ("The Colours of Flowers") an interesting theory of the appearance of the colours of flowers, and it is regarded as probable. He observed that most of the simplest flowers are yellow; the more advanced flowers of simple families, and the simpler flowers of slightly advanced families, are generally white or pink; the most advanced flowers of all families, and almost all the flowers of the more advanced families, are red, purple, or blue; and the most advanced flowers of the most advanced families are always either blue or variegated. Professor Henslow adds a number of equally significant facts with the same tendency, so that we have strong reason to conceive the floral world as passing through successive phases of colour in the Tertiary Era. At first it would be a world of yellows and greens, like that of the Mesozoic vegetation, but brighter. In time splashes of red and white would lie on the face of the landscape; and later would come the purples, the rich blues, and the variegated colours of the more advanced flowers.

Why the colours came at all is a question closely connected with the general story of the evolution of the flower, at which we must glance. The essential characteristic of the flower, in the botanist's judgment, is the central green organ which you find—say, in a lily—standing out in the middle of the floral structure, with a number of yellow-coated rods round it. The yellow rods bear the male germinal elements (pollen); the central pistil encloses the ovules, or female elements. "Angiosperm" means "covered-seed plant," and its characteristic is this protection of the ovules within a special chamber, to which the pollen alone may penetrate. Round these essential organs are the coloured petals of the corolla (the chief part of the flower to the unscientific mind) and the sepals, often also coloured, of the calyx.

There is no doubt that all these parts arose from modifications of the leaves or stems of the primitive plant; though whether the bright leaves of the corolla are directly derived from ordinary leaves, or are enlarged and flattened stamens, has been disputed. And to the question why these bright petals, whose colour and variety of form lend such charm to the world of flowers, have been developed at all, most botanists will give a prompt and very interesting reply. As both male and female elements are usually in one flower, it may fertilise itself, the pollen falling directly on the pistil. But fertilisation is more sure and effective if the pollen comes from a different individual—if there is "cross fertilisation." This may be accomplished by the simple agency of the wind blowing the pollen broadcast, but it is done much better by insects, which brush against the stamens, and carry grains of the pollen to the next flower they visit.

We have here a very fertile line of development among the primitive flowers. The insects begin to visit them, for their pollen or juices, and cross-fertilise them. If this is an advantage, attractiveness to insects will become so important a feature that natural selection will develop it more and more. In plain English, what is meant is that those flowers which are more attractive to insects will be the most surely fertilised and breed most, and the prolonged application of this principle during hundreds of thousands of years will issue in the immense variety of our flowers. They will be enriched with little stores of honey and nectar; not so mysterious an advantage, when we reflect on the concentration of the juices in the neighbourhood of the seed. Then they must "advertise" their stores, and the strong perfumes and bright colours begin to develop, and ensure posterity to their possessors. The shape of the corolla will be altered in hundreds of ways, to accommodate and attract the useful visitor and shut out the mere robber. These utilities, together with the various modifying agencies of different environments, are generally believed to have led to the bewildering variety and great beauty of our floral world.

It is proper to add that this view has been sharply challenged by a number of recent writers. It is questioned if colours and scents do attract insects; though several recent series of experiments seem to show that bees are certainly attracted by colours. It is questioned if cross-fertilisation has really the importance ascribed to it since the days of Darwin. Some of these writers believe that the colours and the peculiar shape which the petals take in some flowers (orchises, for instance) have been evolved to deter browsing animals from eating them. The theory is thus only a different application of natural selection; Professor Henslow, on the other hand, stands alone in denying the selection, and believing that the insects directly developed the scents, honeys, colours, and shapes by mechanical irritation. The great majority of botanists adhere to the older view, and see in the wonderful Tertiary expansion of the flowers a manifold adaptation to the insect friends and insect foes which then became very abundant and varied.

Resisting the temptation to glance at the marvellous adaptations which we find to-day in our plant world— the insect-eating plants, the climbers, the parasites, the sensitive plants, the water-storing plants in dry regions, and so on—we must turn to the consideration of the insects themselves. We have already studied the evolution of the insect in general, and seen its earlier forms. The Tertiary Era not only witnessed a great deployment of the insects, but was singularly rich in means of preserving them. The "fly in amber" has ceased to be a puzzle even to the inexpert. Amber is the resin that exuded from pine-like trees, especially in the Baltic region, in the Eocene and Oligocene periods. Insects stuck in the resin, and were buried under fresh layers of it, and we find them embalmed in it as we pick up the resin on the shores of the Baltic to-day. The Tertiary lakes were also important cemeteries of insects. A great bed at Florissart, in Colorado, is described by one of the American experts who examined it as "a Tertiary Pompeii." It has yielded specimens of about a thousand species of Tertiary insects. Near the large ancient lake, of which it marks the site, was a volcano, and the fine ash yielded from the cone seems to have buried myriads of insects in the water. At Oeningen a similar lake-deposit has, although only a few feet thick, yielded 900 species of insects.

Yet these rich and numerous finds throw little light on the evolution of the insect, except in the general sense that they show species and even genera quite different from those of to-day. No new families of insects have appeared since the Eocene, and the ancient types had by that time disappeared. Since the Eocene, however, the species have been almost entirely changed, so that the insect record, from its commencement in the Primary Era, has the stamp of evolution on every page of it. Unfortunately, insects, especially the higher and later insects, are such frail structures that they are only preserved in very rare conditions. The most important event of the insect-world in the Tertiary is the arrival of the butterflies, which then appear for the first time. We may assume that they spread with great rapidity and abundance in the rich floral world of the mid-Jurassic. More than 13,000 species of Lepidoptera are known to-day, and there are probably twice that number yet to be classified by the entomologist. But so far the Tertiary deposits have yielded only the fragmentary remains of about twenty individual butterflies.

The evolutionary study of the insects is, therefore, not so much concerned with the various modifications of the three pairs of jaws, inherited from the primitive Tracheate, and the wings, which have given us our vast variety of species. It is directed rather to the more interesting questions of what are called the "instincts" of the insects, the remarkable metamorphosis by which the young of the higher orders attain the adult form, and the extraordinary colouring and marking of bees, wasps, and butterflies. Even these questions, however, are so large that only a few words can be said here on the tendencies of recent research.

In regard to the psychic powers of insects it may be said, in the first place, that it is seriously disputed among the modern authorities whether even the highest insects (the ant, bee, and wasp) have any degree whatever of the intelligence which an earlier generation generously bestowed on them. Wasmann and Bethe, two of the leading authorities on ants, take the negative view; Forel claims that they show occasional traces of intelligence. It is at all events clear that the enormous majority of, if not all, their activities—and especially those activities of the ant and the bee which chiefly impress the imagination—are not intelligent, but instinctive actions. And the second point to be noted is that the word "instinct," in the old sense of some innate power or faculty directing the life of an animal, has been struck out of the modern scientific dictionary. The ant or bee inherits a certain mechanism of nerves and muscles which will, in certain circumstances, act in the way we call "instinctive." The problem is to find how this mechanism and its remarkable actions were slowly evolved.

In view of the innumerable and infinitely varied forms of "instinct" in the insect world we must restrict ourselves to a single illustration—say, the social life of the ants and the bees. We are not without indications of the gradual development of this social life. In the case of the ant we find that the Tertiary specimens—and about a hundred species are found in Switzerland alone, whereas there are only fifty species in the whole of Europe to-day—all have wings and are, apparently, of the two sexes, not neutral. This seems to indicate

266

that even in the mid-Tertiary some millions of years after the first appearance of the ant, the social life which we admire in the ants today had not yet been developed. The Tertiary bees, on the other hand, are said to show some traces of the division of labour (and modification of structure) which make the bees so interesting; but in this case the living bees, rising from a solitary life through increasing stages of social co-operation, give us some idea of the gradual development of this remarkable citizenship.

It seems to me that the great selective agency which has brought about these, and many other remarkable activities of the insects (such as the storing of food with their eggs by wasps), was probably the occurrence of periods of cold, and especially the beginning of a winter season in the Cretaceous or Tertiary age. In the periods of luxuriant life (the Carboniferous, the Jurassic, or the Oligocene), when insects swarmed and varied in every direction, some would vary in the direction of a more effective placing of the eggs; and the supervening period of cold and scarcity would favour them. When a regular winter season set in, this tendency would be enormously increased. It is a parallel case to the evolution of the birds and mammals from the reptiles. Those that varied most in the direction of care for the egg and the young would have the largest share in the next generation. When we further reflect that since the Tertiary the insect world has passed through the drastic disturbance of the climate in the great Ice-Age, we seem to have an illuminating clue to one of the most remarkable features of higher insect life.

The origin of the colour marks' and patterns on so many of the higher insects, with which we may join the origin of the stick-insects, leaf-insects, etc., is a subject of lively controversy in science to-day. The protective value of the appearance of insects which look almost exactly like dried twigs or decaying leaves, and of an arrangement of the colours of the wings of butterflies which makes them almost invisible when at rest, is so obvious that natural selection was confidently invoked to explain them. In other cases certain colours or marks seemed to have a value as "warning colours," advertising the nauseousness of their possessors to the bird, which had learned to recognise them; in other cases these colours and marks seemed to be borrowed by palatable species, whose unconscious "mimicry" led to their survival; in other cases, again, the patterns and spots were regarded as "recognition marks," by which the male could find his mate.

Science is just now passing through a phase of acute criticism—as the reader will have realised by this time—and many of the positions confidently adopted in the earlier constructive stage are challenged. This applies to the protective colours, warning colours, mimicry, etc., of insects. Probably some of the affirmations of the older generation of evolutionists were too rigid and extensive; and probably the denials of the new generation are equally exaggerated. When all sound criticism has been met, there remains a vast amount of protective colouring, shaping, and marking in the insect world of which natural selection gives us the one plausible explanation. But the doctrine of natural selection does not mean that every feature of an animal shall have a certain utility. It will destroy animals with injurious variations and favour animals with useful variations; but there may be a large amount of variation, especially in colour, to which it is quite indifferent. In this way much colour-marking may develop, either from ordinary embryonic variations or (as experiment on butterflies shows) from the direct influence of surroundings which has no vital significance. In this way, too, small variations of no selective value may gradually increase until they chance to have a value to the animal. [*]

* For a strong statement of the new critical position see
Dewar and Finn's "Making of Species," 1909, ch. vi.

The origin of the metamorphosis, or pupa-stage, of the higher insects, with all its wonderful protective devices, is so obscure and controverted that we must pass over it. Some authorities think that the sleep-stage has been evolved for the protection of the helpless transforming insect; some believe that it occurs because movement would be injurious to the insect in that stage; some say that the muscular system is actually dissolved in its connections; and some recent experts suggest that it is a reminiscence of the fact that the ancestors of the metamorphosing insects were addicted to internal parasitism in their youth. It is one of the problems of the future. At present we have no fossil pupa-remains (though we have one caterpillar) to guide us. We must leave these fascinating but difficult problems of insect life, and glance at the evolution of the birds.

To the student of nature whose interest is confined to one branch of science the record of life is a mysterious Succession of waves. A comprehensive view of nature, living and non-living, past and present, discovers scores of illuminating connections, and even sees at times the inevitable sequence of events. Thus if the rise of the Angiospermous vegetation on the ruins of the Mesozoic world is understood in the light of geological and climatic changes, and the consequent deploying of the insects, especially the suctorial insects, is a natural result, the simultaneous triumph of the birds is not unintelligible. The grains and fruits of the Angiosperms and the vast swarms of insects provided immense stores of food; the annihilation of the Pterosaurs left a whole stratum of the earth free for their occupation.

We saw that a primitive bird, with very striking reptilian features, was found in the Jurassic rocks, suggesting very clearly the evolution of the bird from the reptile in the cold of the Permian or Triassic period. In the Cretaceous we found the birds distributed in a number of genera, but of two leading types. The Ichthyornis type was a tern-like flying bird, with socketed teeth and biconcave vertebrae like the reptile, but otherwise fully evolved into a bird. Its line is believed to survive in the gannets, cormorants, pelicans, and frigate-birds of to-day. The less numerous Hesperornis group were large and powerful divers. Then there is a blank in the record, representing the Cretaceous upheaval, and it unfortunately conceals the first great ramification of the bird world. When the light falls again on the Eocene period we find great numbers of our familiar types quite developed. Primitive types of gulls, herons, pelicans, quails, ibises, flamingoes, albatrosses, buzzards, hornbills, falcons, eagles, owls, plovers, and woodcocks are found in the Eocene beds; the Oligocene beds add parrots, trogons, cranes, marabouts, secretary-birds, grouse, swallows, and woodpeckers. We cannot suppose that every type has been preserved, but we see that our bird-world was virtually created in the early part of the Tertiary Era.

With these more or less familiar types were large ostrich-like survivors of the older order. In the bed of the sea which covered the site of London in the Eocene are found the remains of a toothed bird (Odontopteryx), though the teeth are merely sharp outgrowths of the edge of the bill. Another bird of the same period and region (Gastornis) stood about ten feet high, and must have looked something like a wading ostrich. Other large waders, even more ostrich-like in structure, lived in North America; and in Patagonia the remains have been found of a massive bird, about eight feet high, with a head larger than that of any living animal except the elephant, rhinoceros, and hippopotamus (Chamberlin).

The absence of early Eocene remains prevents us from tracing the lines of our vast and varied bird-kingdom to their Mesozoic beginnings. And when we appeal to the zoologist to supply the missing links of relationship, by a comparison of the structures of living birds, we receive only uncertain and very general suggestions. [*] He tells us that the ostrich-group (especially the emus and cassowaries) are one of the most primitive stocks of the bird world, and that the ancient Dinornis group and the recently extinct moas seem to be offshoots of that stock. The remaining many thousand species of Carinate birds (or flying birds with a keel [carina]-shaped breast-bone for the attachment of the flying muscles) are then gathered into two great branches, which are "traceable to a common stock" (Pycraft), and branch in their turn along the later lines of development. One of these lines—the pelicans, cormorants, etc.—seems to be a continuation of the Ichthyornis type of the Cretaceous, with the Odontopteryx as an Eocene offshoot; the divers, penguins, grebes, and petrels represent another ancient stock, which may be related to the Hesperornis group of the Cretaceous. Dr. Chalmers Mitchell thinks that the "screamers" of South America are the nearest representatives of the common ancestor of the keel-breasted birds. But even to give the broader divisions of the 19,000 species of living birds would be of little interest to the general reader.

* The best treatment of the subject will be found in W. P.
Pycraft's History of Birds, 1910.

The special problems of bird-evolution are as numerous and unsettled as those of the insects. There is the same dispute as to "protective colours" and "recognition marks", the same uncertainty as to the origin of such instinctive practices as migration and nesting. The general feeling is that the annual migration had its origin in the overcrowding of the regions in which birds could live all the year round. They therefore pushed northward in the spring and remained north until the winter impoverishment drove them south again. On this view each group would be returning to its ancestral home, led by the older birds, in the great migration flights. The curious paths they follow are believed by some authorities to mark the original lines of their spread, preserved from generation to generation through the annual lead of the older birds. If we recollect the Ice-Age which drove the vast majority of the birds south at the end of the Tertiary, and imagine them later following the northward retreat of the ice, from their narrowed and overcrowded southern territory, we may not be far from the secret of the annual migration.

A more important controversy is conducted in regard to the gorgeous plumage and other decorations and weapons of the male birds. Darwin, as is known, advanced a theory of "sexual selection" to explain these. The male peacock, to take a concrete instance, would have developed its beautiful tail because, through tens of thousands of generations, the female selected the more finely tailed male among the various suitors. Dr. Wallace and other authorities always disputed this aesthetic sentiment and choice on the part of the female. The general opinion today is that Darwin's theory could not be sustained in the range and precise sense he gave to it. Some kind of display by the male in the breeding season would be an advantage, but to suppose that the females of any species of birds or mammals had the definite and uniform taste necessary for the creation of male characters by sexual selection is more than difficult. They seem to be connected in origin rather with the higher vitality of the male, but the lines on which they were selected are not yet understood.

This general sketch of the enrichment of the earth with flowering plants, insects, and birds in the Tertiary Era is all that the limits of the present work permit us to give. It is an age of exuberant life and abundant food; the teeming populations overflow their primitive boundaries, and, in adapting themselves to every form of diet, every phase of environment, and every device of capture or escape, the spreading organisms are moulded into tens of thousands of species. We shall see this more clearly in the evolution of the mammals. What we chiefly learn from the present chapter is the vital interconnection of the various parts of nature. Geological changes favour the spread of a certain type of vegetation. Insects are attracted to its nutritious seed-organs, and an age of this form of parasitism leads to a signal modification of the jaws of the insects themselves and to the lavish variety and brilliance of the flowers. Birds are attracted to the nutritious matter enclosing the seeds, and, as it is an advantage to the plant that its seeds be scattered beyond the already populated area, by passing through the alimentary canal of the bird, and being discharged with its excrements, a fresh line of evolution leads to the appearance of the large and coloured fruits. The birds, again, turn upon the swarming insects, and the steady selection they exercise leads to the zigzag flight and the protective colour of the butterfly, the concealment of the grub and the pupa, the marking of the caterpillar, and so on. We can understand the living nature of to-day as the outcome of that teeming, striving, changing world of the Tertiary Era, just as it in turn was the natural outcome of the ages that had gone before.

CHAPTER XVII. THE ORIGIN OF OUR MAMMALS

In our study of the evolution of the plant, the insect, and the bird we were seriously thwarted by the circumstance that their frames, somewhat frail in themselves, were rarely likely to be entombed in good conditions for preservation. Earlier critics of evolution used, when they were imperfectly acquainted with the conditions of fossilisation, to insinuate that this fragmentary nature of the geological record was a very convenient refuge for the evolutionist who was pressed for positive evidence. The complaint is no longer found in any serious work. Where we find excellent conditions for preservation, and animals suitable for preservation living in the midst of them, the record is quite satisfactory. We saw how the chalk has yielded remains of sea-urchins in the actual and gradual process of evolution. Tertiary beds which represent the muddy bottoms of tranquil lakes are sometimes equally instructive in their fossils, especially of shell-fish. The Paludina of a certain Slavonian lake-deposit is a classical example. It changes so greatly in the successive levels of the deposit that, if the intermediate forms were not preserved, we should divide it into several different species. The Planorbis is another well-known example. In this case we have a species evolving along several distinct lines into forms which differ remarkably from each other.

The Tertiary mammals, living generally on the land and only coming by accident into deposits suitable for preservation, cannot be expected to reveal anything like this sensible advance from form to form. They were, however, so numerous in the mid-Tertiary, and their bones are so well calculated to survive when they do fall into suitable conditions, that we can follow their development much more easily than that of the birds. We find a number of strange patriarchal beasts entering the scene in the early Eocene, and spreading into a great variety of forms in the genial conditions of the Oligocene and Miocene. As some of these forms advance, we begin to descry in them the features, remote and shadowy at first, of the horse, the deer, the elephant, the whale, the tiger, and our other familiar mammals. In some instances we can trace the evolution with a wonderful fullness, considering the remoteness of the period and the conditions of preservation. Then, one by one, the abortive, the inelastic, the ill-fitted types are destroyed by changing conditions or powerful carnivores, and the field is left to the mammals which filled it when man in turn began his destructive career.

The first point of interest is the origin of these Tertiary mammals. Their distinctive advantage over the mammals of the Mesozoic Era was the possession by the mother of a placenta (the "after-birth" of the higher mammals), or structure in the womb by which the blood-vessels of the mother are brought into such association with those of the foetus that her blood passes into its arteries, and it is fully developed within the warm shelter of her womb. The mammals of the Mesozoic had been small and primitive animals, rarely larger than a rat, and never rising above the marsupial stage in organisation. They not only continued to exist, and give rise to their modern representatives (the opossum, etc.) during the Tertiary Era, but they shared the general prosperity. In Australia, where they were protected from the higher carnivorous mammals, they gave rise to huge elephant-like wombats (Diprotodon), with skulls two or three feet in length. Over the earth generally, however, they were superseded by the placental mammals, which suddenly break into the geological record in the early Tertiary, and spread with great vigour and rapidity over the four continents.

Were they a progressive offshoot from the Mesozoic Marsupials, or Monotremes, or do they represent a separate stock from the primitive half-reptile and half-mammal family? The point is disputed; nor does the scantiness of the record permit us to tell the place of their origin. The placental structure would be so great an advantage in a cold and unfavourable environment that some writers look to the northern land, connecting Europe and America, for their development. We saw, however, that this northern region was singularly warm until long after the spread of the mammals. Other experts, impressed by the parallel development of the mammals and the flowering plants, look to the elevated parts of eastern North America.

Such evidence as there is seems rather to suggest that South Africa was the cradle of the placental mammals. We shall find that many of our mammals originated in Africa; there, too, is found to-day the most primitive representative of the Tertiary mammals, the hyrax; and there we find in especial abundance the remains of the mammal-like reptiles (Theromorphs) which are regarded as their progenitors. Further search in the unexplored geological treasures and dense forests of Africa is needed. We may provisionally conceive the placental mammals as a group of the South African early mammals which developed a fortunate variation in womb-structure during the severe conditions of the early Mesozoic. In this new structure they would have no preponderant advantage as long as the genial Jurassic age favoured the great reptiles, and they may have remained as small and insignificant as the Marsupials. But with the fresh upheaval and climatic disturbance at the end of the Jurassic, and during the Cretaceous, they spread northward, and replaced the dying reptiles, as the Angiosperms replaced the dying cycads. When they met the spread of the Angiosperm vegetation they would receive another great stimulus to development.

They appear in Europe and North America in the earliest Cretaceous. The rise of the land had connected many hitherto isolated regions, and they seem to have poured over every bridge into all parts of the four continents. The obscurity of their origin is richly compensated by their intense evolutionary interest from the moment they enter the geological record. We have seen this in the case of every important group of plants and animals, and can easily understand it. The ancestral group was small and local; the descendants are widely spread. While, therefore, we discover remains of the later phases of development in our casual cuttings and quarries, the ancestral tomb may remain for ages in some unexplored province of the geological world. If this region is, as we suspect, in

Africa, our failure to discover it as yet is all the more intelligible.

But these mammals of the early Tertiary are still of such a patriarchal or ancestral character that the student of evolution can dispense with their earlier phase. They combine in their primitive frames, in an elementary way, the features which we now find distributed in widely removed groups of their descendants. Most of them fall into two large orders: the Condylarthra, the ancestral herbivores from which we shall find our horses, oxen, deer, elephants, and hogs gradually issuing, and the Creodonta, the patriarchal carnivores, which will give birth to our lions and tigers, wolves and foxes, and their various cousins. As yet even the two general types of herbivore and carnivore are so imperfectly separated that it is not always possible to distinguish between them. Nearly all of them have the five-toed foot of the reptile ancestor; and the flat nails on their toes are the common material out of which the hoof of the ungulate and the claw of the carnivore will be presently fashioned. Nearly all have forty-four simply constructed teeth, from which will be evolved the grinders and tusks of the elephant or the canines of the tiger. They answer in every respect to the theory that some primitive local group was the common source of all our great mammals. With them are ancestral forms of Edentates (sloths, etc.) and Insectivores (moles, etc.), side-branches developing according to their special habits; and before the end of the Eocene we find primitive Rodents (squirrels, etc.) and Cheiroptera (bats).

From the description of the Tertiary world which we have seen in the last chapter we understand the rapid evolution of the herbivorous Condylarthra. The rich vegetation which spreads over the northern continents, to which they have penetrated, gives them an enormous vitality and fecundity, and they break into groups, as they increase in number, adapted to the different conditions of forest, marsh, or grass-covered plain. Some of them, swelling lazily on the abundant food, and secure for a time in their strength, become the Deinosaurs of their age, mere feeding and breeding machines. They are massive, sluggish, small-brained animals, their strong stumpy limbs terminating in broad five-toed feet. Coryphodon, sometimes as large as an ox, is a typical representative. It is a type fitted only for prosperous days, and these Amblypoda, as they are called, will disappear as soon as the great carnivores are developed.

Another doomed race, or abortive experiment of early mammal life, were the remarkable Deinocerata ("terrible-horned" mammals). They sometimes measured thirteen feet in length, but had little use for brain in the conditions in which they were developed. The brain of the Deinoceras was only one-eighth the size of the brain of a rhinoceros of the same bulk; and the rhinoceros is a poor-brained representative of the modern mammals. To meet the growing perils of their race they seem to have developed three pairs of horns on their long, flat skulls, as we find on them three pairs of protuberances. A late specimen of the group, Tinoceras, had a head four feet in length, armed with these six horns, and its canine teeth were developed into tusks sometimes seven or eight inches in length. They suggest a race of powerful but clumsy and grotesque monsters, making a last stand, and developing such means of protection as their inelastic nature permitted. But the horns seem to have proved a futile protection against the advancing carnivores, and the race was extinguished. The horns may, of course, have been mainly developed by, or for, the mutual butting of the males.

The extinction of these races will remind many readers of a theory on which it is advisable to say a word. It will be remembered that the last of the Deinosaurs and the Ammonites also exhibited some remarkable developments in their last days. These facts have suggested to some writers the idea that expiring races pass through a death-agony, and seem to die a natural death of old age like individuals. The Trilobites are quoted as another instance; and some ingenious writers add the supposed eccentricities of the Roman Empire in its senile decay and a number of other equally unsubstantial illustrations.

There is not the least ground for this fantastic speculation. The destruction of these "doomed races" is as clearly traceable to external causes as is the destruction of the Roman Empire; nor, in fact, did the Roman Empire develop any such eccentricities as are imagined in this superficial theory. What seem to our eye the "eccentricities" and "convulsions" of the Ceratopsia and Deinocerata are much more likely to be defensive developments against a growing peril, but they were as futile against the new carnivores as were the assegais of the Zulus against the European. On the other hand, the eccentricities of many of the later Trilobites—the LATEST Trilobites, it may be noted, were chaste and sober specimens of their race, like the last Roman patricians—and of the Ammonites may very well have been caused by physical and chemical changes in the sea-water. We know from experiment that such changes have a disturbing influence, especially on the development of eggs and larvae; and we know from the geological record that such changes occurred in the periods when the Trilobites and Ammonites perished. In fine, the vast majority of extinct races passed through no "convulsions" whatever. We may conclude that races do not die; they are killed.

The extinction of these races of the early Condylarthra, and the survival of those races whose descendants share the earth with us to-day, are quite intelligible. The hand of natural selection lay heavy on the Tertiary herbivores. Apart from overpopulation, forcing groups to adapt themselves to different regions and diets, and apart from the geological disturbances and climatic changes which occurred in nearly every period, the shadow of the advancing carnivores was upon them. Primitive but formidable tigers, wolves, and hyenas were multiplying, and a great selective struggle set in. Some groups shrank from the battle by burrowing underground like the rabbit; some, like the squirrel or the ape, took refuge in the trees; some, like the whale and seal, returned to the water; some shrank into armour, like the armadillo, or behind fences of spines, like the hedgehog; some, like the bat, escaped into the air. Social life also was probably developed at this time, and the great herds had their sentinels and leaders. But the most useful qualities of the large vegetarians, which lived on grass and leaf, were acuteness of perception to see the danger, and speed of limb to escape it. In other words, increase of brain and sense-power and increase of speed were the primary requisites. The clumsy early Condylarthra failed to meet the tests, and perished; the other branches of the race were more plastic, and, under the pressure of a formidable enemy, were gradually moulded into the horse, the deer, the ox, the antelope, and the elephant.

We can follow the evolution of our mammals of this branch most easily by studying the modification of the feet and limbs. In a running attitude—the experiment may be tried—the weight of the body is shifted from the flat sole of the foot, and thrown upon the toes, especially the central toes. This indicates the line of development of the Ungulates (hoofed animals) in the struggle of the Tertiary Era. In the early Eocene we find the Condylarthra (such as Phenacodus) with flat five-toed feet, and such a mixed combination of characters that they "might serve very well for the ancestors of all the later Ungulata" (Woodward). We then presently find this generalised Ungulate branching into three types, one of which seems to be a patriarchal tapir, the second is regarded as a very remote ancestor of the horse, and the third foreshadows the rhinoceros. The feet have now only three or four toes; one or two of the side-toes have disappeared. This evolution, however, follows two distinct lines. In one group of these primitive Ungulates the main axis of the limb, or the stress of the weight, passes through the middle toe.

This group becomes the Perissodactyla ("odd-toed" Ungulates) of the zoologist, throwing out side-branches in the tapir and the rhinoceros, and culminating in the one-toed horse. In the other line, the Artiodactyla (the "even-toed" or cloven-hoofed Ungulates), the main axis or stress passes between the third and fourth toes, and the group branches into our deer, oxen, sheep, pigs, camels, giraffes, and hippopotamuses. The elephant has developed along a separate and very distinctive line, as we shall see, and the hyrax is a primitive survivor of the ancestral group.

Thus the evolutionist is able to trace a very natural order in the immense variety of our Ungulates. He can follow them in theory as they slowly evolve from their primitive Eocene ancestor according to their various habits and environments; he has a very rich collection of fossil remains illustrating the stages of their development; and in the hyrax (or "coney") he has one more of those living fossils, or primitive survivors, which still fairly preserve the ancestral form. The hyrax has four toes on the front foot and three on the hind foot, and the feet are flat. Its front teeth resemble those of a rodent, and its molars those of the rhinoceros. In many respects it is a most primitive and generalised little animal, preserving the ancestral form more or less faithfully since Tertiary days in the shelter of the African Continent.

The rest of the Ungulates continued to develop through the Tertiary, and fortunately we are enabled to follow the development of two of the most interesting of them, the horse and the elephant, in considerable detail. As I said above, the primitive Ungulate soon branches into three types which dimly foreshadow the tapir, the horse, and the rhinoceros, the three forms of the Perissodactyl. The second of these types is the Hyracotherium. It has no distinct equine features, and is known only from the skull, but the authorities regard it as the progenitor (or representative of the progenitors) of the horse-types. In size it must have been something like the rabbit or the hyrax. Still early in the Eocene, however, we find the remains of a small animal (Eohippus), about the size of a fox, which is described as "undoubtedly horse-like." It had only three toes on its hind feet, and four on its front feet; though it had also a splint-bone, representing the shrunken and discarded fifth toe, on its fore feet. Another form of the same period (Protorohippus) shows the central of the three toes on the hind foot much enlarged, and the lateral toes shrinking. The teeth, and the bones and joints of the limbs, are also developing in the direction of the horse.

In the succeeding geological period, the Oligocene, we find several horse-types in which the adaptation of the limbs to running on the firm grassy plains and of the teeth to eating the grass continues. Mesohippus has lost the fourth toe of the fore foot, which is now reduced to a splintbone, and the lateral toes of its hind foot are shrinking. In the Miocene period there is a great development of the horse-like mammals. We have the remains of more than forty species, some continuing the main line of development on the firm and growing prairies of the Miocene, some branching into the softer meadows or the forests, and giving rise to types which will not outlive the Tertiary. They have three toes on each foot, and have generally lost even the rudimentary trace of the fourth toe. In most of them, moreover, the lateral toes—except in the marsh-dwelling species, with spreading feet—scarcely touch the ground, while the central toe is developing a strong hoof. The leg-bones are longer, and have a new type of joint; the muscles are concentrated near the body. The front teeth are now chopping incisors, and the grinding teeth approach those of the modern horse in the distribution of the enamel, dentine, and cement. They are now about the size of a donkey, and must have had a distinctly horsy appearance, with their long necks and heads and tapering limbs. One of them, Merychippus, was probably in the direct line of the evolution of the horse. From Hipparion some of the authorities believe that the zebras may have been developed. Miohippus, Protohippus, and Hypohippus, varying in size from that of a sheep to that of a donkey, are other branches of this spreading family.

In the Pliocene period the evolution of the main stem culminates in the appearance of the horse, and the collateral branches are destroyed. Pliohippus is a further intermediate form. It has only one toe on each foot, with two large splint bones, but its hoof is less round than that of the horse, and it differs in the shape of the skull and the length of the teeth. The true horse (Equus) at length appears, in Europe and America, before the close of the Tertiary period. As is well known, it still has the rudimentary traces of its second and fourth toes in the shape of splint bones, and these bones are not only more definitely toe-shaped in the foal before birth, but are occasionally developed and give us a three-toed horse.

From these successive remains we can confidently picture the evolution, during two or three million years, of one of our most familiar mammals. It must not, of course, be supposed that these fossil remains all represent "ancestors of the horse." In some cases they may very well do so; in others, as we saw, they represent sidebranches of the family which have become extinct. But even such successive forms as the Eohippus, Mesohippus, Miohippus, and Pliohippus must not be arranged in a direct line as the pedigree of the horse. The family became most extensive in the Miocene, and we must regard the casual fossil specimens we have discovered as illustrations of the various phases in the development of the horse from the primitive Ungulate. When we recollect what we saw in an earlier chapter about the evolution of grassy plains and the successive rises of the land during the Tertiary period, and when we reflect on the simultaneous advance of the carnivores, we can without difficulty realise this evolution of our familiar companion from a hyrax-like little animal of two million years ago.

We have not in many cases so rich a collection of intermediate forms as in the case of the horse, but our fossil mammals are numerous enough to suggest a similar development of all the mammals of to-day. The primitive family which gave birth to the horse also gave us, as we saw, the tapir and the rhinoceros. We find ancestral tapirs in Europe and America during the Tertiary period, but the later cold has driven them to the warm swamps of Brazil and Malaysia. The rhinoceros has had a long and interesting history. From the primitive Hyrochinus of the Eocene, in which it is dimly foreshadowed, we pass to a large and varied family in the later periods. In the Oligocene it spreads into three great branches, adapted, respectively, to life on the elevated lands, the lowlands, and the water. The upland type (Hyracodon) was a light-limbed running animal, well illustrating the close relation to the horse. The aquatic representative (Metamynodon) was a stumpy and bulky animal. The intermediate lowland type was probably the ancestor of the modern animal. All three forms were yet hornless. In the Miocene the lowland type (Leptaceratherium, Aceratherium, etc.) develops vigorously, while the other branches die. The European types now have two horns, and in one of the American species (Diceratherium) we see a commencement of the horny growths from the skull. We shall see later that the rhinoceros continued in Europe even during the severe conditions of the glacial period, in a branch that developed a woolly coat.

There were also in the early Tertiary several sidebranches of the horse-tapir-rhinoceros family. The Palaeotheres were more or less between the horse and the tapir in structure; the Anoplotheres between the tapir and the ruminant. A third doomed branch, the Titanotheres, flourished vigorously for a time, and begot some strange and monstrous forms (Brontops, Titanops, etc.). In the larger specimens the body was about fourteen feet long, and stood ten feet from the ground. The long, low skull had a pair of horns over the snout. They perished like the equally powerful but equally sluggish and stupid Deinocerata. The Tertiary was an age of brain rather than of brawn. As compared with their early Tertiary representatives' some of our modern mammals have increased seven or eight-fold in brain-capacity.

While the horses and tapirs and rhinoceroses were being gradually evolved from the primitive types, the Artiodactyl branch of the Ungulates—the pigs, deer, oxen, etc.—were also developing. We must dismiss them briefly. We saw that the primitive herbivores divided early in the Eocene into the "odd-toed" and "even-toed" varieties; the name refers, it will be remembered, not to the number of toes, but to the axis of stress. The Artiodactyl group must have quickly branched in turn, as we find very primitive hogs and camels before the end of the Eocene. The first hog-like creature (Homacodon) was much smaller than the hog of to-day, and had strong canine teeth, but in the Oligocene the family gave rise to a large and numerous race, the Elotheres. These "giant-pigs," as they have been called, with two toes on each foot, flourished vigorously for a time in Europe and America, but were extinguished in the Miocene, when the true pigs made their appearance. Another doomed race of the time is represented by the Hyopotamus, an animal between the pig and the hippopotamus; and the Oreodontids, between the hog and the deer, were another unsuccessful branch of the early race. The hippopotamus itself was widespread in Europe, and a familiar form in the rivers of Britain, in the latter part of the Tertiary.

The camel seems to be traceable to a group of primitive North American Ungulates (Paebrotherium, etc.) in the later Eocene period. The Paebrotherium, a small animal about two feet long, is followed by Pliauchenia, which points toward the llamas and vicunas, and Procamelus, which clearly foreshadows the true camel. In the Pliocene the one branch went southward, to develop into the llamas and vicunas, and the other branch crossed to Asia, to develop into the camels. Since that time they have had no descendants in North America.

The primitive giraffe appears suddenly in the later Tertiary deposits of Europe and Asia. The evidence points to an invasion from Africa, and, as the region of development is unknown and unexplored, the evolution of the giraffe remains a matter of speculation. Chevrotains flourished in Europe and North America in the Oligocene, and are still very primitive in structure, combining features of the hog and the ruminants. Primitive deer and oxen begin in the Miocene, and seem to have an earlier representative in certain American animals (Protoceras), of which the male has a pair of blunt outgrowths between the ears. The first true deer are hornless (like the primitive muskdeer of Asia to-day), but by the middle of the Miocene the males have small two-pronged antlers, and as the period proceeds three or four more prongs are added. It is some confirmation of the evolutionary embryonic law that we find the antlers developing in this way in the individual stag to-day. A very curious race of ruminants in the later Tertiary was a large antelope (Sivatherium) with four horns. It had not only the dimensions, but apparently some of the characters, of an elephant.

The elephant itself, the last type of the Ungulates, has a clearer line of developments. A chance discovery of fossils in the Fayum district in Egypt led Dr. C. W. Andrews to make a special exploration, and on the remains which he found he has constructed a remarkable story of the evolution of the elephant. [*] It is clear that the elephant was developed in Africa, and a sufficiently complete series of remains has been found to give a good idea of the origin of its most distinctive features. In the Eocene period there lived in the Egyptian region an animal, something like the tapir in size and appearance, which had its second incisors developed into small tusks and—to judge from the nasal opening in the skull—a somewhat prolonged snout. This animal (Moeritherium) only differed from the ordinary primitive Ungulate in these incipient elephantine features. In the later Eocene a larger and more advanced animal, the Palaeomastodon, makes its appearance. Its tusks are larger (five or six inches long), its molars more elephantine, the air-cells at the back of the head more developed. It would look like a small elephant, except that it had a long snout, instead of a flexible trunk, and a projecting lower jaw on which the snout rested.

*See this short account, "Guide to the Elephants in the
British Museum," 1908.

Up to the beginning of the Miocene, Africa was, as we saw, cut off from Europe and Asia by the sea which stretched from Spain to India. Then the land rose, and the elephant passed by the new tracts into the north. Its next representative, Tetrabelodon, is found in Asia and Europe, as well as North Africa. The frame is as large as that of a medium-sized elephant, and the increase of the air-cells at the back of the skull shows that an increased weight has to be sustained by the muscles of the neck. The nostrils are shifted further back. The tusks are from twenty to thirty inches long, and round, and only differ from those of the elephant in curving slightly downward, The chin projects as far as the tusks. The neck is shorter and thicker, and, as the animal increases in height, we can understand that the long snout—possibly prehensile at its lower end—is necessary for the animal to reach the ground. But the snout still lies on the projecting lower jaw, and is not a trunk. Passing over the many collateral branches, which diverge in various directions, we next kind that the chin is shortening (in Tetrabelodon longirostris), and, through a long series of discovered intermediate forms, we trace the evolution of the elephant from the mastodon. The long supporting skin disappears, and the enormous snout becomes a flexible trunk. Southern Asia seems to have been the province of this final transformation, and we have remains of some of these primitive elephants with tusks nine and a half feet long. A later species, which wandered over Central and Southern Europe before the close of the Tertiary, stood fifteen feet high at the shoulder, while the mammoth, which superseded it in the days of early man, had at times tusks more than ten feet in length.

It is interesting to reflect that this light on the evolution of one of our most specialised mammals is due to the chance opening of the soil in an obscure

African region. It suggests to us that as geological exploration is extended, many similar discoveries may be made. The slenderness of the geological record is a defect that the future may considerably modify.

From this summary review of the evolution of the Ungulates we must now pass to an even briefer account of the evolution of the Carnivores. The evidence is less abundant, but the characters of the Carnivores consist so obviously of adaptations to their habits and diet that we have little difficulty in imagining their evolution. Their early Eocene ancestors, the Creodonts, gave rise in the Eocene to forms which we may regard as the forerunners of the cat-family and dog-family, to which most of our familiar Carnivores belong. Patriofelis, the "patriarchal cat," about five or six feet in length (without the tail), curiously combines the features of the cat and the seal-family. Cyonodon has a wolf-like appearance, and Amphicyon rather suggests the fox. Primitive weasels, civets, and hyaenas appear also in the Eocene. The various branches of the Carnivore family are already roughly represented, but it is an age of close relationships and generalised characters.

In the Miocene we find the various groups diverging still further from each other and from the extinct stocks. Definite wolves and foxes abound in America, and the bear, civet, and hyaena are represented in Europe, together with vague otter-like forms. The dog-family seems to have developed chiefly in North America. As in the case of the Ungulates, we find many strange side-branches which flourished for a time, but are unknown to-day. Machoerodus, usually known as "the sabre-toothed tiger," though not a tiger, was one of the most formidable of these transitory races. Its upper canine teeth (the "sabres") were several inches in length, and it had enormously distensible jaws to make them effective. The great development of such animals, with large numbers of hyaenas, civets, wolves, bears, and other Carnivores, in the middle and later Tertiary was probably the most effective agency in the evolution of the horse and deer and the extinction of the more sluggish races. The aquatic branch of the Carnivores (seals, walruses, etc.) is little represented in the Tertiary record. We saw, however, that the most primitive representatives of the elephant-stock had also some characters of the seal, and it is thought that the two had a common origin.

The Moeritherium was a marsh-animal, and may very well have been cousin to the branch of the family which pushed on to the seas, and developed its fore limbs into paddles.

The Rodents are represented in primitive form early in the Eocene period. The teeth are just beginning to show the characteristic modification for gnawing. A large branch of the family, the Tillodonts, attained some importance a little later. They are described as combining the head and claws of a bear with the teeth of a rodent and the general characters of an ungulate. In the Oligocene we find primitive squirrels, beavers, rabbits, and mice. The Insectivores also developed some of the present types at an early date, and have since proved so unprogressive that some regard them as the stock from which all the placental mammals have arisen.

The Cetacea (whales, porpoises, etc.) are already represented in the Eocene by a primitive whale-like animal (Zeuglodon) of unknown origin. Some specimens of it are seventy feet in length. It has large teeth, sometimes six inches long, and is clearly a terrestrial mammal that has returned to the waters. Some forms even of the modern whale develop rudimentary teeth, and in all forms the bony structure of the fore limbs and degenerate relic of a pelvis and back limbs plainly tell of the terrestrial origin. Dolphins appear in the Miocene.

Finally, the Edentates (sloths, anteaters, and armadilloes) are represented in a very primitive form in the early Eocene. They are then barely distinguishable from the Condylarthra and Creodonta, and seem only recently to have issued from a common ancestor with those groups. In the course of the Tertiary we find them—especially in South America, which was cut off from the North and its invading Carnivores during the Eocene and Miocene—developed into large sloths, armadilloes, and anteaters. The reconnection with North America in the Pliocene allowed the northern animals to descend, but gigantic sloths (Megatherium) and armadilloes (Glyptodon) flourished long afterwards in South America. The Megatherium attained a length of eighteen feet in one specimen discovered, and the Glyptodon often had a dorsal shield (like that of the armadillo) from six to eight feet long, and, in addition, a stoutly armoured tail several feet long.

The richness and rapidity of the mammalian development in the Tertiary, of which this condensed survey will convey some impression, make it impossible to do more here than glance over the vast field and indicate the better-known connections. It will be seen that evolution not only introduces a lucid order and arrangement into our thousands of species of living and fossil mammals, but throws an admirable light on the higher animal world of our time. The various orders into which the zoologist puts our mammals are seen to be the branches of a living tree, approaching more and more closely to each other in early Tertiary times, in spite of the imperfectness of the geological record. We at last trace these diverging lines to a few very primitive, generalised, patriarchal groups, which in turn approach each other very closely in structure, and plainly suggest a common Cretaceous ancestor. Whether that common ancestor was an Edentate, an Insectivore, or Creodont, or something more primitive than them all, is disputed. But the divergence of nearly all the lines of our mammal world from those patriarchal types is admirably clear. In the mutual struggle of carnivore and herbivore, in adaptation to a hundred different environments (the water, the land, and the air, the tree, the open plain, the underground, the marsh, etc.) and forms of diet, we find the descendants of these patriarchal animals gradually developing their distinctive characters. Then we find the destructive agencies of living and inorganic nature blotting out type after type, and the living things that spread over the land in the later Tertiary are found to be broadly identical with the living things of to-day. The last great selection, the northern Ice-Age, will give the last touches of modernisation.

CHAPTER XVIII. THE EVOLUTION OF MAN

We have reserved for a closer inquiry that order of the placental mammals to which we ourselves belong, and on which zoologists have bestowed the very proper and distinguishing name of the Primates. Since the days of Darwin there has been some tendency to resent the term "lower animals," which man applies to his poorer relations. But, though there is no such thing as an absolute standard by which we may judge the "higher" or "lower" status of animals or plants, the extraordinary power which man has by his brain development attained over both animate and inanimate nature fully justifies the phrase. The Primate order is, therefore, of supreme interest as the family that gave birth to man, and it is important to discover the agencies which impelled some primitive member of it to enter upon the path which led to this summit of organic nature.

The order includes the femurs, a large and primitive family with ape-like features—the Germans call them "half-apes"—the monkeys, the man-like apes, and man. This classification according to structure corresponds with the successive appearance of the various families in the geological record. The femurs appear in the Eocene; the monkeys, and afterwards the apes, in the Miocene, the first semi-human forms in the Pleistocene, though they must have been developed before this. It is hardly necessary to say that science does not regard man as a descendant of the known anthropoid apes, or these as descended from the monkeys. They are successive types or phases of development, diverging early from each other. Just as the succeeding horse-types of the record are not necessarily related to each other in a direct line, yet illustrate the evolution of a type which culminates in the horse, so the spreading and branching members of the Primate group illustrate the evolution of a type of organism which culminates in man. The particular relationship of the various families, living and dead, will need careful study.

That there is a general blood-relationship, and that man is much more closely related to the anthropoid apes than to any of the lower Primates, is no longer a matter of controversy. In Rudolph Virchow there died, a few years ago, the last authoritative man of science to express any doubt about it. There are, however, non-scientific writers who, by repeating the ambiguous phrase that it is "only a theory," convey the impression to inexpert readers that it is still more or less an open question. We will therefore indicate a few of the lines of evidence which have overcome the last hesitations of scientific men, and closed the discussion as to the fact.

The very close analogy of structure between man and the ape at once suggests that they had a common ancestor. There are cases in which two widely removed animals may develop a similar organ independently, but there is assuredly no possibility of their being alike in all organs, unless by common inheritance. Yet the essential identity of structure in man and the ape is only confirmed by every advance of science, and

298

would of itself prove the common parentage. Such minor differences as there are between man and the higher ape—in the development of the cerebrum, the number of the teeth or ribs, the distribution of the hair, and so on—are quite explicable when we reflect that the two groups must have diverged from each other more than a million years ago.

Examining the structure of man more closely, we find this strong suggestion of relationship greatly confirmed. It is now well known that the human body contains a number of vestigial "organs"—organs of no actual use, and only intelligible as vestiges of organs that were once useful. Whatever view we take of the origin of man, each organ in his frame must have a meaning; and, as these organs are vestigial and useless even in the lowest tribes of men, who represent primitive man, they must be vestiges of organs that were of use in a remote pre-human ancestor. The one fact that the ape has the same vestigial organs as man would, on a scientific standard of evidence, prove the common descent of the two. But these interesting organs themselves point back far earlier than a mixed ape-human ancestor in many cases.

The shell of cartilage which covers the entrance to the ear—the gristly appendage which is popularly called the ear—is one of the clearest and most easily recognised of these organs. The "ear" of a horse or a cat is an upright mobile shell for catching the waves of sound. The human ear has the appearance of being the shrunken relic of such an organ, and, when we remove the skin, and find seven generally useless muscles attached to it, obviously intended to pull the shell in all

299

directions (as in the horse), there can be no doubt that the external ear is a discarded organ, a useless legacy from an earlier ancestor. In cases where it has been cut off it was found that the sense of hearing was scarcely, if at all, affected. Now we know that it is similarly useless in all tribes of men, and must therefore come from a pre-human ancestor. It is also vestigial in the higher apes, and it is only when we descend to the lower monkeys and femurs that we see it approaching its primitive useful form. One may almost say that it is a reminiscence of the far-off period when, probably in the early Tertiary, the ancestors of the Primates took to the trees. The animals living on the plain needed acute senses to detect the approach of their prey or their enemies; the tree-dweller found less demand on his sense of hearing, the "speaking-trumpet" was discarded, and the development of the internal ear proceeded on the higher line of the perception of musical sounds.

We might take a very large number of parts of the actual human body, and discover that they are similar historical or archaeological monuments surviving in a modern system, but we have space only for a few of the more conspicuous.

The hair on the body is a vestigial organ, of actual use to no race of men, an evident relic of the thick warm coat of an earlier ancestor. It in turn recalls the dwellers in the primeval forest. In most cases—not all, because the wearing of clothes for ages has modified this feature—it will be found that the hairs on the arm tend upward from the wrist to the elbow, and downward from the shoulder to the elbow. This very peculiar feature becomes intelligible when we find that some of the apes also have it, and that it has a certain use in their case. They put their hands over their heads as they sit in the trees during ram, and in that position the sloping hair acts somewhat like the thatched roof of a cottage.

Again, it will be found that in the natural position of standing we are not perfectly flat-footed, but tend to press much more on the outer than on the inner edge of the foot. This tendency, surviving after ages of living on the level ground, is a lingering effect of the far-off arboreal days.

A more curious reminiscence is seen in the fact that the very young infant, flabby and powerless as it is in most of its muscles, is so strong in the muscles of the hand and arm that it can hang on to a stick by its hands, and sustain the whole weight of its body, for several minutes. Finally, our vestigial tail—for we have a tail comparable to that of the higher apes—must be mentioned. In embryonic development the tail is much longer than the legs, and some children are born with a real tail, which they move as the puppy does, according to their emotional condition. Other features of the body point back to an even earlier stage. The vermiform appendage—in which some recent medical writers have vainly endeavoured to find a utility—is the shrunken remainder of a large and normal intestine of a remote ancestor. This interpretation of it would stand even if it were found to have a certain use in the human body. Vestigial organs are sometimes pressed into a secondary use when their original function has been lost. The danger of this appendage in the human body to-day is due to the fact that it is a blind alley leading off the alimentary canal, and has a very narrow opening. In the ape the opening is larger, and, significantly enough, it is still larger in the human foetus. When we examine some of the lower mammals we discover the meaning of it. It is in them an additional storage chamber in the alimentary system. It is believed that a change to a more digestible diet has made this additional chamber superfluous in the Primates, and the system is slowly suppressing it.

Other reminiscences of this earlier phase are found in the many vestigial muscles which are found in the body to-day. The head of the quadruped hangs forward, and is held by powerful muscles and ligaments in the neck. We still have the shrunken remainder of this arrangement. Other vestigial muscles are found in the forehead, the scalp, the nose—many people can twitch the nostrils and the scalp—and under the skin in many parts of the body. These are enfeebled remnants of the muscular coat by which the quadruped twitches its skin, and drives insects away. A less obvious feature is found by the anatomist in certain blood-vessels of the trunk. As the blood flows vertically in a biped and horizontally in a quadruped, the arrangement of the valves in the blood-vessels should be different in the two cases; but it is the same in us as in the quadruped. Another trace of the quadruped ancestor is found in the baby. It walks "on all fours" so long, not merely from weakness of the limbs, but because it has the spine of a quadruped.

A much more interesting fact, but one less easy to interpret, is that the human male has, like the male ape, organs for suckling the young. That there are real milk-glands, usually vestigial, underneath the teats in the breast of the boy or the man is proved by the many known cases in which men have suckled the young. Several friends of the present writer have seen this done in India and Ceylon by male "wet-nurses." As there is no tribe of men or species of ape in which the male suckles the young normally, we seem to be thrown back once more upon an earlier ancestor. The difficulty is that we know of no mammal of which both parents suckle the young, and some authorities think that the breasts have been transferred to the male by a kind of embryonic muddle. That is difficult to believe, as no other feature has ever been similarly transferred to the opposite sex. In any case the male breasts are vestigial organs. Another peculiarity of the mammary system is that sometimes three, four, or five pairs of breasts appear in a woman (and several have been known even in a man). This is, apparently, an occasional reminiscence of an early mammal ancestor which had large litters of young and several pairs of breasts.

But there are features of the human body which recall an ancestor even earlier than the quadruped. The most conspicuous of these is the little fleshy pad at the inner corner of each eye. It is a common feature in mammals, and is always useless. When, however, we look lower down in the animal scale we find that fishes and reptiles (and birds) have a third eyelid, which is drawn across the eye from this corner. There is little room to doubt that the little fleshy vestige in the mammal's eye is the shrunken remainder of the lateral eyelid of a remote fish-ancestor.

A similar reminiscence is found in the pineal body, a small and useless object, about the size and shape of a hazel-nut, in the centre of the brain. When we examine the reptile we find a third eye in the top of the head. The skin has closed over it, but the skull is still, in many cases, perforated as it is for the eyes in front. I have seen it standing out like a ball on the head of a dead crocodile, and in the living tuatara—the very primitive New Zealand lizard—it still has a retina and optic nerve. As the only animal in nature to-day with an eye in this position (the Pyrosome, a little marine animal of the sea-squirt family) is not in the line of reptile and mammal ancestry, it is difficult to locate the third eye definitely. But when we find the skin closing over it in the amphibian and reptile, then the bone, and then see it gradually atrophying and being buried under the growing brain, we must refer it to some early fish-ancestor. This ancestor, we may recall, is also reflected for a time in the gill-slits and arches, with their corresponding fish-like heart and blood-vessels, during man's embryonic development, as we saw in a former chapter.

These are only a few of the more conspicuous instances of vestigial structures in man. Metchnikoff describes about a hundred of them. Even if there were no remains of primitive man pointing in the direction of a common ancestry with the ape, no lower types of men in existence with the same tendency, no apes found in nature to-day with a structure so strikingly similar to that of man, and no fossil records telling of the divergence of forms from primitive groups in past time, we should be forced to postulate the evolution of man in order to explain his actual features. The vestigial structures must be interpreted as we interpret the buttons on the back of a man's coat. They are useless reminiscences of an age in which they were useful. When their witness to the past is supported by so many converging lines of evidence it becomes irresistible. I will add only one further testimony which has been brought into court in recent years.

The blood consists of cells, or minute disk-shaped corpuscles, floating in a watery fluid, or serum. It was found a few years ago, in the course of certain experiments in mixing the blood of animals, that the serum of one animal's blood sometimes destroyed the cells of the other animal's blood, and at other times did not. When the experiments were multiplied, it was found that the amount of destructive action exercised by one specimen of blood upon another depended on the nearness or remoteness of relationship between the animals. If the two are closely related, there is no disturbance when their blood is mixed; when they are not closely related, the serum of one destroys the cells of the other, and the intensity of the action is in proportion to their remoteness from each other. Another and more elaborate form of the experiment was devised, and the law was confirmed. On both tests it was found by experiment that the blood of man and of the anthropoid ape behaved in such a way as to prove that they were closely related. The blood of the monkey showed a less close relationship—a little more remote in the New World than in the Old World monkeys; and the blood of the femur showed a faint and distant relationship.

The FACT of the evolution of man and the apes from a common ancestor is, therefore, outside the range of controversy in science; we are concerned only to retrace the stages of that evolution, and the agencies which controlled it. Here, unfortunately, the geological record gives us little aid. Tree-dwelling animals are amongst the least likely to be buried in deposits which may preserve their bones for ages. The distribution of femur and ape remains shows that the order of the Primates has been widespread and numerous since the middle of the Tertiary Era, yet singularly few remains of the various families have been preserved.

Hence the origin of the Primates is obscure. They are first foreshadowed in certain femur-like forms of the Eocene period, which are said in some cases (Adapis) to combine the characters of pachyderms and femurs, and in others (Anaptomorphus) to unite the features of Insectivores and femurs. Perhaps the more common opinion is that they were evolved from a branch of the Insectivores, but the evidence is too slender to justify an opinion. It was an age when the primitive placental mammals were just beginning to diverge from each other, and had still many features in common. For the present all we can say is that in the earliest spread of the patriarchal mammal race one branch adopted arboreal life, and evolved in the direction of the femurs and the apes. The generally arboreal character of the Primates justifies this conclusion.

In the Miocene period we find a great expansion of the monkeys. These in turn enter the scene quite suddenly, and the authorities are reduced to uncertain and contradictory conjectures as to their origin. Some think that they develop not from the femurs, but along an independent line from the Insectivores, or other ancestors of the Primates. We will not linger over these early monkeys, nor engage upon the hopeless task of tracing their gradual ramification into the numerous families of the present age. It is clear only that they soon divided into two main streams, one of which spread into the monkeys of America and the other into the monkeys of the Old World. There are important anatomical differences between the two. The monkeys remained in Central and Southern Europe until near the end of the Tertiary. Gradually we perceive that the advancing cold is driving them further south, and the monkeys of Gibraltar to-day are the diminished remnant of the great family that had previously wandered as far as Britain and France.

A third wave, also spreading in the Miocene, equally obscure in its connection with the preceding, introduces the man-like apes to the geologist. Primitive gibbons (Pliopithecus and Pliobylobates), primitive chimpanzees (Palaeopithecus), and other early anthropoid apes (Oreopithecus, Dryopithecus, etc.), lived in the trees of Southern Europe in the second part of the Tertiary Era. They are clearly disconnected individuals of a large and flourishing family, but from the half-dozen specimens we have yet discovered no conclusion can be drawn, except that the family is already branching into the types of anthropoid apes which are familiar to us.

Of man himself we have no certain and indisputable trace in the Tertiary Era. Some remains found in Java of an ape-man (Pithecanthropus), which we will study later, are now generally believed, after a special investigation on the spot, to belong to the Pleistocene period. Yet no authority on the subject doubts that the human species was evolved in the Tertiary Era, and very many, if not most, of the authorities believe that we have definite proof of his presence. The early story of mankind is gathered, not so much from the few fragments of human remains we have, but from the stone implements which were shaped by his primitive intelligence and remain, almost imperishable, in the soil over which he wandered. The more primitive man was, the more ambiguous would be the traces of his shaping of these stone implements, and the earliest specimens are bound to be a matter of controversy. It is claimed by many distinguished authorities that flints slightly touched by the hand of man, or at least used as implements by man, are found in abundance in England, France, and Germany, and belong to the Pliocene period. Continental authorities even refer some of them to the Miocene and the last part of the Oligocene.

The question whether an implement-using animal, which nearly all would agree to regard as in some degree human, wandered over what is now the South of England (Kent, Essex, Dorsetshire, etc.) as many hundred thousand years ago as this claim would imply, is certainly one of great interest. But there would be little use in discussing here the question of the "Eoliths," as these disputed implements are called. A very keen controversy is still being conducted in regard to them, and some of the highest authorities in England, France, and Germany deny that they show any trace of human workmanship or usage. Although they have the support of such high authorities as Sir J. Prestwich, Sir E. Ray Lankester, Lord Avebury, Dr. Keane, Dr. Blackmore, Professor Schwartz, etc., they are one of those controverted testimonies on which it would be ill-advised to rely in such a work as this.

We must say, then, that we have no undisputed traces of man in the Tertiary Era. The Tertiary implements which have been at various times claimed in France, Italy, and Portugal are equally disputed; the remains which were some years ago claimed as Tertiary in the United States are generally disallowed; and the recent claims from South America are under discussion. Yet it is the general feeling of anthropologists that man was evolved in the Tertiary Era. On the one hand, the anthropoid apes were highly developed by the Miocene period, and it would be almost incredible that the future human stock should linger hundreds of thousands of years behind them. On the other hand, when we find the first traces of man in the Pleistocene, this development has already proceeded so far that its earlier phase evidently goes back into the Tertiary. Let us pass beyond the Tertiary Era for a moment, and examine the earliest and most primitive remains we have of human or semi-human beings.

The first appearance of man in the chronicle of terrestrial life is a matter of great importance and interest. Even the least scientific of readers stands, so to say, on tiptoe to catch a first glimpse of the earliest known representative of our race, and half a century of discussion of evolution has engendered a very wide interest in the early history of man. [*]

* A personal experience may not be without interest in this
connection. Among the many inquiries directed to me in
regard to evolution I received, in one month, a letter from
a negro in British Guiana and an extremely sensible query
from an inmate of an English asylum for the insane! The
problem that beset the latter of the two was whether the
Lemuranda preceded the Lemurogona in Eocene times. He had
found a contradiction in the statements of two scientific
writers.

Fortunately, although these patriarchal bones are very scanty—two teeth, a thigh-bone, and the skull-cap—we are now in a position to form some idea of the nature of their living owner. They have been subjected to so searching a scrutiny and discussion since they were found in Java in 1891 and 1892 that there is now a general agreement as to their nature. At first some of the experts thought that they were the remains of an abnormally low man, and others that they belonged to an abnormally high ape. The majority held from the start that they belonged to a member of a race almost midway between the highest family of apes and the lowest known tribe of men, and therefore fully merited the name of "Ape-Man" (Pithecanthropus). This is now the general view of anthropologists.

The Ape-Man of Java was in every respect entitled to that name. The teeth suggest a lower part of the face in which the teeth and lips projected more than in the most ape-like types of Central Africa. The skull-cap has very heavy ridges over the eyes and a low receding forehead, far less human than in any previously known prehistoric skull. The thigh-bone is very much heavier than any known human femur of the same length, and so appreciably curved that the owner was evidently in a condition of transition from the semi-quadrupedal crouch of the ape to the erect attitude of man. The Ape-Man, in other words, was a heavy, squat, powerful, bestial-looking animal; of small stature, but above the pygmy standard; erect in posture, but with clear traces of the proneness of his ancestor; far removed from the highest ape in brainpower, but almost equally far removed from the lowest savage that is known to us. We shall see later that there is some recent criticism, by weighty authorities, of the earlier statements in regard to the brain of primitive man. This does not apply to the Ape-Man of Java. The average cranial capacity (the amount of brain-matter the skull may contain) of the chimpanzees, the highest apes, is about 600 cubic centimetres. The average cranial capacity of the lowest races of men, of moderate stature, is about 1200. And the cranial capacity of Ape-Man was about 900

It is immaterial whether or no these bones belong to the same individual. If they do not, we have remains of two or three individuals of the same intermediate species. Nor does it matter whether or no this early race is a direct ancestor of the later races of men, or an extinct offshoot from the advancing human stock. It is, in either case, an illustration of the intermediate phase between the ape and man The more important tasks are to trace the relationship of this early human stock to the apes, and to discover the causes of its superior evolution.

The first question has a predominantly technical interest, and the authorities are not agreed in replying to it. We saw that, on the blood-test, man showed a very close relationship to the anthropoid apes, a less close affinity to the Old World monkeys, a more remote affinity to the American monkeys, and a very faint and distant affinity to the femurs. A comparison of their structures suggests the same conclusion. It is, therefore, generally believed that the anthropoid apes and man had a common ancestor in the early Miocene or Oligocene, that this group was closely related to the ancestral group of the Old World monkeys, and that all originally sprang from a primitive and generalised femur-group. In other words, a branch of the earliest femur-like forms diverges, before the specific femur-characters are fixed, in the direction of the monkey; in this still vague and patriarchal group a branch diverges, before the monkey-features are fixed, in the direction of the anthropoids; and this group in turn spreads into a number of types, some of which are the extinct apes of the Miocene, four become the gorilla, chimpanzee, orang, and gibbon of to-day, and one is the group that will become man. To put it still more precisely, if we found a whole series of remains of man's ancestors during the Tertiary, we should probably class them, broadly, as femur-remains in the Eocene, monkey-remains in the Oligocene, and ape-remains in the

Miocene. In that sense only man "descends from a monkey."

The far more important question is: How did this one particular group of anthropoid animals of the Miocene come to surpass all its cousins, and all the rest of the mammals, in brain-development? Let us first rid the question of its supposed elements of mystery and make of it a simple problem. Some imagine that a sudden and mysterious rise in intelligence lifted the progenitor of man above its fellows. The facts very quickly dispel this illusion. We may at least assume that the ancestor of man was on a level with the anthropoid ape in the Miocene period, and we know from their skulls that the apes were as advanced then as they are now. But from the early Miocene to the Pleistocene is a stretch of about a million years on the very lowest estimate. In other words, man occupied about a million years in travelling from the level of the chimpanzee to a level below that of the crudest savage ever discovered. If we set aside the Java man, as a possible survivor of an earlier phase, we should still have to say that, much more than a million years after his departure from the chimpanzee level, man had merely advanced far enough to chip stone implements; because we find no other trace whatever of intelligence than this until near the close of the Palaeolithic period. If there is any mystery, it is in the slowness of man's development.

Let us further recollect that it is a common occurrence in the calendar of life for a particular organ to be especially developed in one member of a particular group more than in the others. The trunk of the elephant, the neck of the giraffe, the limbs of the horse or deer, the canines of the satire-toothed tiger, the wings of the bat, the colouring of the tiger, the horns of the deer, are so many examples in the mammal world alone. The brain is a useful organ like any other, and it is easy to conceive that the circumstances of one group may select it just as the environment of another group may lead to the selection of speed, weapons, or colouring. In fact, as we saw, there was so great and general an evolution of brain in the Tertiary Era that our modern mammals quite commonly have many times the brain of their Tertiary ancestors. Can we suggest any reasons why brain should be especially developed in the apes, and more particularly still in the ancestors of man?

The Primate group generally is a race of tree-climbers. The appearance of fruit on early Tertiary trees and the multiplication of carnivores explain this. The Primate is, except in a few robust cases, a particularly defenceless animal. When its earliest ancestors came in contact with fruit and nut-bearing trees, they developed climbing power and other means of defence and offense were sacrificed. Keenness of scent and range of hearing would now be of less moment, but sight would be stimulated, especially when soft-footed climbing carnivores came on the scene. There is, however, a much deeper significance in the adoption of climbing, and we must borrow a page from the modern physiology of the brain to understand it.

The stress laid in the modern education of young children on the use of the hands is not merely due to a feeling that they should handle objects as well as read

314

about them. It is partly due to the belief of many distinguished physiologists that the training of the hands has a direct stimulating effect on the thought-centres in the brain. The centre in the cerebrum which controls the use of the hands is on the fringe of the region which seems to be concerned in mental operations. For reasons which will appear presently, we may add that the centres for controlling the muscles of the face and head are in the same region. Any finer training or the use of the hands will develop the centre for the fore limbs, and, on the principles, may react on the more important region of the cortex. Hence in turning the fore foot into a hand, for climbing and grasping purposes, the primitive Primate entered upon the path of brain-development. Even the earliest Primates show large brains in comparison with the small brains of their contemporaries.

It is a familiar fact in the animal world that when a certain group enters upon a particular path of evolution, some members of the group advance only a little way along it, some go farther, and some outstrip all the others. The development of social life among the bees will illustrate this. Hence we need not be puzzled by the fact that the lemurs have remained at one mental level, the monkeys at another, and the apes at a third. It is the common experience of life; and it is especially clear among the various races of men. A group becomes fitted to its environment, and, as long as its surroundings do not change, it does not advance. A related group, in a different environment, receives a particular stimulation, and advances. If, moreover, a group remains unstimulated for ages, it may become so rigid in its type that it loses the capacity to advance. It is generally believed that the lowest races of men, and even some of the higher races like the Australian aboriginals, are in this condition. We may expect this "unteachability" in a far more stubborn degree in the anthropoid apes, which have been adapted to an unchanging environment for a million years.

All that we need further suppose is—and it is one of the commonest episodes in terrestrial life—that one branch of the Miocene anthropoids, which were spread over a large part of the earth, received some stimulus to change which its cousins did not experience. It is sometimes suggested that social life was the great advantage which led to the superior development of mind in man. But such evidence as there is would lead us to suppose that primitive man was solitary, not social. The anthropoid apes are not social, but live in families, and are very unprogressive. On the other hand, the earliest remains of prehistoric man give no indication of social life. Fire-places, workshops, caves, etc., enter the story in a later phase. Some authorities on prehistoric man hold very strongly that during the greater part of the Old Stone Age (two-thirds, at least, of the human period) man wandered only in the company of his mate and children. [*]

* The point will be more fully discussed later. This account
   of prehistoric life is well seen in Mortillet's
   Prehistorique (1900). The lowest races also have no tribal
   life, and Professor Westermarck is of opinion that early man
   was not social.

We seem to have the most plausible explanation of the divergence of man from his anthropoid cousins in the fact that he left the trees of his and their ancestors. This theory has the advantage of being a fact—for the Ape-Man race of Java has already left the trees—and providing a strong ground for brain-advance. A dozen reasons might be imagined for his quitting the trees—migration, for instance, to a region in which food was more abundant, and carnivores less formidable, on the ground-level—but we will be content with the fact that he did. Such a change would lead to a more consistent adoption of the upright attitude, which is partly found in the anthropoid apes, especially the gibbons. The fore limb would be no longer a support of the body; the hand would be used more for grasping; and the hand-centre in the brain would be proportionately stimulated. The adoption of the erect attitude would further lead to a special development of the muscles of the head and face, the centre for which is in the same important region in the cortex. There would also be a direct stimulation of the brain, as, having neither weapons nor speed, the animal would rely all the more on sight and mind. If we further suppose that this primitive being extended the range of his hunting, from insects and small or dead birds to small land-animals, the stimulation would be all the greater. In a word, the very fact of a change from the trees to the ground suggests a line of brain-development which may plausibly be conceived, in the course of a million years, to evolve an Ape-Man out of a man-like ape. And we are not introducing any imaginary factor in this view of human origins.

The problem of the evolution of man is often approached in a frame of mind not far removed from that of the educated, but inexpert, European who stands before the lowly figure of the chimpanzee, and wonders by what miracle the gulf between it and himself was bridged. That is to lay a superfluous strain on the imagination. The proper term of comparison is the lowest type of human being known to us, since the higher types of living men have confessedly evolved from the lower. But even the lowest type of existing or recent savage is not the lowest level of humanity. Whether or no the Tasmanian or the Yahgan is a primitive remnant of the Old Stone Age, we have a far lower depth in the Java race. What we have first to do is to explain the advance to that level, in the course of many hundreds of thousands of years: a period fully a hundred times as long as the whole history of civilisation. Time itself is no factor in evolution, but in this case it is a significant condition. It means that, on this view of the evolution of man, we are merely assuming that an advance in brain-development took place between the Miocene and the Pleistocene, not similar to, but immeasurably less than, the advance which we know to have been made in the last fifty thousand years. In point of fact, the most mysterious feature of the evolution of man was its slowness. We shall see that, to meet the facts, we must suppose man to have made little or no progress during most of this vast period, and then to have received some new stimulation to develop. What it was we have now to inquire.

CHAPTER XIX. MAN AND THE GREAT ICE-AGE

In discussing the development of plants and animals during the Tertiary Era we have already perceived the shadow of the approaching Ice-Age. We found that in the course of the Tertiary the types which were more sensitive to cold gradually receded southward, and before its close Europe, Asia, and North America presented a distinctly temperate aspect. This is but the penumbra of the eclipse. When we pass the limits of the Tertiary Era, and enter the Quaternary, the refrigeration steadily proceeds, and, from temperate, the aspect of much of Europe and North America becomes arctic. From six to eight million square miles of the northern hemisphere are buried under fields of snow and ice, and even in the southern regions smaller glacial sheets spread from the foot of the higher ranges of mountains.

It is unnecessary to-day to explain at any length the evidences by which geologists trace this enormous glaciation of the northern hemisphere. There are a few works still in circulation in which popular writers, relying on the obstinacy of a few older geologists, speak lightly of the "nightmare" of the Ice-Age. But the age has gone by in which it could seriously be suggested that the boulders strewn along the east of Scotland—fragments of rock whose home we must seek in Scandinavia—were brought by the vikings as ballast for their ships. Even the more serious controversy, whether the scratches and the boulders which we find on the face of Northern Europe and America were due to floating or land ice, is virtually settled. Several decades of research have detected the unmistakable signs of glacial action over this vast area of the northern hemisphere. Most of Europe north of the Thames and the Danube, nearly all Canada and a very large part of the United States, and a somewhat less expanse of Northern Asia, bear to this day the deep scars of the thick, moving ice-sheets. Exposed rock-surfaces are ground and scratched, beds of pebbles are twisted and contorted hollows are scooped out, and moraines—the rubbish-heaps of the glaciers— are found on every side. There is now not the least doubt that, where the great Deinosaurs had floundered in semi-tropical swamps, where the figs and magnolias had later flourished, where the most industrious and prosperous hives of men are found to-day, there was, in the Pleistocene period, a country to which no parallel can be found outside the polar circles to-day.

The great revolution begins with the gathering of snows on the mountains. The Alps and Pyrenees had now, we saw, reached their full stature, and the gathering snows on their summits began to glide down toward the plains in rivers of ice. The Apennines (and even the mountains of Corsica), the Balkans, Carpathians, Caucasus, and Ural Mountains, shone in similar mantles of ice and snow. The mountains of Wales, the north of England, Scotland, and Scandinavia had even heavier burdens, and, as the period advanced, their sluggish streams of ice poured slowly over the plains. The trees struggled against the increasing cold in the narrowing tracts of green; the animals died, migrated to the south, or put on arctic coats. At length the ice-sheets of Scandinavia met the spreading sheets from Scotland and Wales, and crept over Russia and Germany, and an almost continuous mantle, from which only a few large areas of arctic vegetation peeped out, was thrown over the greater part of Europe. Ten thousand feet thick where it left the hills of Norway and Sweden, several thousand feet thick even in Scotland, the ice-sheet that resulted from the fusion of the glaciers gradually thinned as it went south, and ended in an irregular fringe across Central Europe. The continent at that time stretched westward beyond the Hebrides and some two hundred miles beyond Ireland. The ice-front followed this curve, casting icebergs into the Atlantic, then probably advanced up what is now the Bristol Channel, and ran across England and Europe, in a broken line, from Bristol to Poland. South of this line there were smaller ice-fields round the higher mountains, north of it almost the whole country presented the appearance that we find in Greenland to-day.

In North America the glaciation was even more extensive. About four million square miles of the present temperate zone were buried under ice and snow. From Greenland, Labrador, and the higher Canadian mountains the glaciers poured south, until, in the east, the mass of ice penetrated as far as the valley of the Mississippi. The great lakes of North America are permanent memorials of its Ice-Age, and over more than half the country we trace the imprint and the relics of the sheet. South America, Australia, Tasmania, and New Zealand had their glaciated areas. North Asia was largely glaciated, but the range of the ice-sheet is not yet determined in that continent.

This summary statement will convey some idea of the extraordinary phase through which the earth passed in the early part of the present geological era. But it must be added that a singular circumstance prolonged the glacial regime in the northern hemisphere. Modern geologists speak rather of a series of successive ice-sheets than of one definite Ice-Age. Some, indeed, speak of a series of Ice-Ages, but we need not discuss the verbal question. It is now beyond question that the ice-sheet advanced and retreated several times during the Glacial Epoch. The American and some English geologists distinguished six ice-sheets, with five intermediate periods of more temperate climate. The German and many English and French geologists distinguish four sheets and three interglacial epochs. The exact number does not concern us, but the repeated spread of the ice is a point of some importance. The various sheets differed considerably in extent. The wide range of the ice which I have described represents the greatest extension of the glaciation, and probably corresponds to the second or third of the six advances in Dr. Geikie's (and the American) classification.

Before we consider the biological effect of this great of refrigeration of the globe, we must endeavour to understand the occurrence itself. Here we enter a world of controversy, but a few suggestions at least may be gathered from the large literature of the subject, which dispel much of the mystery of the Great Ice-Age.

It was at one time customary to look out beyond the earth itself for the ultimate causes of this glaciation. Imagine the sheet of ice, which now spreads widely round the North Pole, shifted to another position on the surface of the planet, and you have a simple explanation of the occurrence. In other words, if we suppose that the axis of the earth does not consistently point in one direction—that the great ball does not always present the same average angle in relation to the sun—the poles will not always be where they are at present, and the Pleistocene Ice-Age may represent a time when the north pole was in the latitude of North Europe and North America. This opinion had to be abandoned. We have no trace whatever of such a constant shifting of the polar regions as it supposes, and, especially, we have no trace that the warm zone correspondingly shifted in the Pleistocene.

A much more elaborate theory was advanced by Dr. Croll, and is still entertained by many. The path of the earth round the sun is not circular, but elliptical, and there are times when the gravitational pull of the other planets increases the eccentricity of the orbit. It was assumed that there are periods of great length, separated from each other by still longer periods, when this eccentricity of the orbit is greatly exaggerated. The effect would be to prolong the winter and shorten the summer of each hemisphere in turn. The total amount of heat received would not alter, but there would be a long winter with less heat per hour, and a short summer with more heat. The short summer would not suffice to melt the enormous winter accumulations of ice and snow, and an ice-age would result. To this theory, again, it is objected that we do not find the regular succession of ice-ages in the story of the earth which the theory demands, and that there is no evidence of an alternation of the ice between the northern and southern hemispheres.

More recent writers have appealed to the sun itself, and supposed that some prolonged veiling of its photosphere greatly reduced the amount of heat emitted by it. More recently still it has been suggested that an accumulation of cosmic or meteoric dust in our atmosphere, or between us and the sun, had, for a prolonged period, the effect of a colossal "fire-screen." Neither of these suppositions would explain the localisation of the ice. In any case we need not have recourse to purely speculative accidents in the world beyond until it is clear that there were no changes in the earth itself which afford some explanation.

This is by no means clear. Some writers appeal to changes in the ocean currents. It is certain that a change in the course of the cold and warm currents of the ocean to-day might cause very extensive changes of climate, but there seems to be some confusion of ideas in suggesting that this might have had an equal, or even greater, influence in former times. Our ocean currents differ so much in temperature because the earth is now divided into very pronounced zones of climate. These zones did not exist before the Pliocene period, and it is not at all clear that any redistribution of currents in earlier times could have had such remarkable consequences. The same difficulty applies to wind-currents.

On the other hand, we have already, in discussing the Permian glaciation, discovered two agencies which are very effective in lowering the temperature of the earth. One is the rise of the land; the other is the thinning of the atmosphere. These are closely related agencies, and we found them acting in conjunction to bring about the Permian Ice-Age. Do we find them at work in the Pleistocene?

It is not disputed that there was a very considerable upheaval of the land, especially in Europe and North America, at the end of the Tertiary Era. Every mountain chain advanced, and our Alps, Pyrenees, Himalaya, etc., attained, for the first time, their present, or an even greater elevation. The most critical geologists admit that Europe, as a whole, rose 4000 feet above its earlier level. Such an elevation would be bound to involve a great lowering of the temperature. The geniality of the Oligocene period was due, like that of the earlier warm periods, to the low-lying land and very extensive water-surface. These conditions were revolutionised before the end of the Tertiary. Great mountains towered into the snow-line, and vast areas were elevated which had formerly been sea or swamp.

This rise of the land involved a great decrease in the proportion of moisture in the atmosphere. The sea surface was enormously lessened, and the mountains would now condense the moisture into snow or cloud to a vastly greater extent than had ever been known before There would also be a more active circulation of the atmosphere, the moist warm winds rushing upward towards the colder elevations and parting with their vapour. As the proportion of moisture in the atmosphere lessened the surface-heat would escape more freely into space, the general temperature would fall, and the evaporation—or production of moisture would be checked, while the condensation would continue. The prolonging of such conditions during a geological period can be understood to have caused the accumulation of fields of snow and ice in the higher regions. It seems further probable that these conditions would lead to a very considerable formation of fog and cloud, and under this protecting canopy the glaciers would creep further down toward the plains.

We have then to consider the possibility of a reduction of the quantity of carbon-dioxide in the atmosphere The inexpert reader probably has a very exaggerated idea of the fall in temperature that would be required to give Europe an Ice-Age. If our average temperature fell about 5-8 degrees C. below the average temperature of our time it would suffice; and it is further calculated that if the quantity of carbon-dioxide in our atmosphere were reduced by half, we should have this required fall in temperature. So great a reduction would not be necessary in view of the other refrigerating agencies. Now it is quite certain that the proportion of carbon-dioxide was greatly reduced in the Pleistocene. The forests of the Tertiary Era would steadily reduce it, but the extensive upheaval of the land at its close would be even more important. The newly exposed surfaces would absorb great quantities of carbon. The ocean, also, as it became colder, would absorb larger and larger quantities of carbon-dioxide. Thus the Pleistocene atmosphere, gradually relieved of its vapours and carbon-dioxide, would no longer retain the heat at the surface. We may add that the growth of reflective surfaces—ice, snow, cloud, etc.—would further lessen the amount of heat received from the sun.

Here, then, we have a series of closely related causes and effects which would go far toward explaining, if they do not wholly suffice to explain, the general fall of the earth's temperature. The basic cause is the upheaval of the land—a fact which is beyond controversy, the other agencies are very plain and recognisable consequences of the upheaval. There are, however, many geologists who do not think this explanation adequate.

It is pointed out, in the first place, that the glaciation seems to have come long after the elevation. The difficulty does not seem to be insurmountable. The reduction of the atmospheric vapour would be a gradual process, beginning with the later part of the elevation and culminating long afterwards. The reduction of the carbon-dioxide would be even more gradual. It is impossible to say how long it would take these processes to reach a very effective stage, but it is equally impossible to show that the interval between the upheaval and the glaciation is greater than the theory demands.

It is also said that we cannot on these principles understand the repeated advance and retreat of the ice-sheet.

This objection, again, seems to fail. It is an established fact that the land sank very considerably during the Ice-Age, and has risen again since the ice disappeared. We find that the crust in places sank so low that an arctic ocean bathed the slopes of some of the Welsh mountains; and American geologists say that their land has risen in places from 2000 to 3000 feet (Chamberlin) since the burden of ice was lifted from it. Here we have the possibility of an explanation of the advances and retreats of the glaciers. The refrigerating agencies would proceed until an enormous burden of ice was laid on the land of the northern hemisphere. The land apparently sank under the burden, the ice and snow melted at the lower level and there was a temperate interglacial period. But the land, relieved of its burden, rose once more, the exposed surface absorbed further quantities of carbon, and a fresh period of refrigeration opened. This oscillation might continue until the two sets of opposing forces were adjusted, and the crust reached a condition of comparative stability.

Finally, and this is the more serious difficulty, it is said that we cannot in this way explain the localisation of the glacial sheets. Why should Europe and North America in particular suffer so markedly from a general thinning of the atmosphere? The simplest answer is to suggest that they especially shared the rise of the land. Geology is not in a position either to prove or disprove this, and it remains only a speculative interpretation of the fact We know at least that there was a great uprise of land in Europe and North America in the Pliocene and Pleistocene and may leave the precise determination of the point to a later age. At the same time other local causes are not excluded.

There may have been a large extension of the area of atmospheric depression which we have in the region of Greenland to-day.

When we turn to the question of chronology we have the same acute difference of opinion as we have found in regard to all questions of geological time. It used to be urged, on astronomical grounds, that the Ice-Age began about 240,000 years ago, and ended about 60,000 years ago, but the astronomical theory is, as I said, generally abandoned. Geologists, on the other hand, find it difficult to give even approximate figures. Reviewing the various methods of calculation, Professor Chamberlin concludes that the time of the first spread of the ice-sheet is quite unknown, the second and greatest extension of the glaciation may have been between 300,000 and a million years ago, and the last ice-extension from 20,000 to 60,000 years ago; but he himself attaches "very little value" to the figures. The chief ice-age was some hundreds of thousands of years ago, that is all we can say with any confidence.

In dismissing the question of climate, however, we should note that a very serious problem remains unsolved. As far as present evidence goes we seem to be free to hold that the ice-ages which have at long intervals invaded the chronicle of the earth were due to rises of the land. Upheaval is the one constant and clearly recognisable feature associated with, or preceding, ice-ages. We saw this in the case of the Cambrian, Permian, Eocene, and Pleistocene periods of cold, and may add that there are traces of a rise of mountains before the glaciation of which we find traces in the middle of the Archaean Era. There are problems still to be solved in connection with each of these very important ages, but in the rise of the land and consequent thinning of the atmosphere we seem to have a general clue to their occurrence. Apart from these special periods of cold, however, we have seen that there has been, in recent geological times, a progressive cooling of the earth, which we have not explained. Winter seems now to be a permanent feature of the earth's life, and polar caps are another recent, and apparently permanent, acquisition. I find no plausible reason assigned for this.

The suggestion that the disk of the sun is appreciably smaller since Tertiary days is absurd; and the idea that the earth has only recently ceased to allow its internal heat to leak through the crust is hardly more plausible. The cause remains to be discovered.

We turn now to consider the effect of the great Ice-Age, and the relation of man to it. The Permian revolution, to which the Pleistocene Ice-Age comes nearest in importance, wrought such devastation that the overwhelming majority of living things perished. Do we find a similar destruction of life, and selection of higher types, after the Pleistocene perturbation? In particular, had it any appreciable effect upon the human species?

A full description of the effect of the great Ice-Age would occupy a volume. The modern landscape in Europe and North America was very largely carved and modelled by the ice-sheet and the floods that ensued upon its melting. Hills were rounded, valleys carved, lakes formed, gravels and soils distributed, as we find them to-day. In its vegetal aspect, also, as we saw, the modern landscape was determined by the Pleistocene revolution. A great scythe slowly passed over the land. When the ice and snow had ended, and the trees and flowers, crowded in the southern area, slowly spread once more over the virgin soil, it was only the temperate species that could pass the zone guarded by the Alps and the Pyrenees. On the Alps themselves the Pleistocene population still lingers, their successful adaptation to the cold now preventing them from descending to the plains.

The animal world in turn was winnowed by the Pleistocene episode. The hippopotamus, crocodile, turtle, flamingo, and other warm-loving animals were banished to the warm zone. The mammoth and the rhinoceros met the cold by developing woolly coats, but the disappearance of the ice, which had tempted them to this departure, seems to have ended their fitness. Other animals which became adapted to the cold—arctic bears, foxes, seals, etc.—have retreated north with the ice, as the sheet melted. For hundreds of thousands of years Europe and North America, with their alternating glacial and interglacial periods, witnessed extraordinary changes and minglings of their animal population. At one time the reindeer, the mammoth, and the glutton penetrate down to the Mediterranean, in the next phase the elephant and hippopotamus again advance nearly to Central Europe. It is impossible here to attempt to unravel these successive changes and migrations. Great numbers of species were destroyed, and at length, when the climatic condition of the earth reached a state of comparative stability, the surviving animals settled in the geographical regions in which we find them to-day.

The only question into which we may enter with any fullness is that of the relation of human development to this grave perturbation of the condition of the globe. The problem is sometimes wrongly conceived. The chief point to be determined is not whether man did or did not precede the Ice-Age. As it is the general belief that he was evolved in the Tertiary, it is clear that he existed in some part of the earth before the Ice-Age. Whether he had already penetrated as far north as Britain and Belgium is an interesting point, but not one of great importance. We may, therefore, refrain from discussing at any length those disputed crude stone implements (Eoliths) which, in the opinion of many, prove his presence in northern regions before the close of the Tertiary. We may also now disregard the remains of the Java Ape-Man. There are authorities, such as Deniker, who hold that even the latest research shows these remains to be Pliocene, but it is disputed. The Java race may be a surviving remnant of an earlier phase of human evolution.

The most interesting subject for inquiry is the fortune of our human and prehuman forerunners during the Pliocene and Pleistocene periods. It may seem that if we set aside the disputable evidence of the Eoliths and the Java remains we can say nothing whatever on this subject. In reality a fact of very great interest can be established. It can be shown that the progress made during this enormous lapse of time—at least a million years—was remarkably slow. Instead of supposing that some extraordinary evolution took place in that

332

conveniently obscure past, to which we can find no parallel within known times, it is precisely the reverse. The advance that has taken place within the historical period is far greater, comparatively to the span of time, than that which took place in the past.

To make this interesting fact clearer we must attempt to measure the progress made in the Pliocene and Pleistocene. We may assume that the precursor of man had arrived at the anthropoid-ape level by the middle of the Miocene period. He is not at all likely to have been behind the anthropoid apes, and we saw that they were well developed in the mid-Tertiary. Now we have a good knowledge of man as he was in the later stage of the Ice-Age—at least a million years later—and may thus institute a useful comparison and form some idea of the advance made.

In the later stages of the Pleistocene a race of men lived in Europe of whom we have a number of skulls and skeletons, besides vast numbers of stone implements. It is usually known as the Neanderthal race, as the first skeleton was found, in 1856, at Neanderthal, near Dusseldorf. Further skeletons were found at Spy, in Belgium, and Krapina, in Croatia. A skull formerly found at Gibraltar is now assigned to the same race. In the last five years a jaw of the same (or an earlier) age has been found at Mauer, near Heidelberg, and several skeletons have been found in France (La Vezere and Chapelle-aux-Saints). From these, and a few earlier fragments, we have a confident knowledge of the features of this early human race.

The highest appreciation of the Neanderthal man—a somewhat flattering appreciation, as we shall see—is that he had reached the level of the Australian black of to-day. The massive frontal ridges over his eyes, the very low, retreating forehead, the throwing of the mass of the brain toward the back of the head, the outthrust of the teeth and jaws, and the complete absence (in some cases) or very slight development of the chin, combine to give the head what the leading authorities call a "bestial" or "simian" aspect. The frame is heavy, powerful, and of moderate height (usually from two to four inches over five feet). The thigh-bones are much more curved than in modern man. We cannot enter here into finer anatomical details, but all the features are consistent and indicate a stage in the evolution from ape-man to savage man.

One point only calls for closer inquiry. Until a year or two ago it was customary to state that in cranial capacity also—that is to say, in the volume of brain-matter that the skull might contain—the Neanderthal race was intermediate between the Ape-Man and modern man. We saw above that the cranial capacity of the highest ape is about 600 cubic centimetres, and that of the Ape-Man (variously given as 850 and 950) is about 900. It was then added that the capacity of the Neanderthal race was about 1200, and that of civilised man (on the average) 1600. This seemed to be an effective and convincing indication of evolution, but recent writers have seriously criticised it. Sir Edwin Ray Lankester, Professor Sollas, and Dr. Keith have claimed in recent publications that the brain of Neanderthal man was as large as, if not larger than, that of modern man. [*] Professor Sollas even observes that "the brain increases in volume as we go backward." This is, apparently, so serious a reversal of the familiar statement in regard to the evolution of man that we must consider it carefully.

*See especially an address by Professor Sollas in the
    Quarterly Journal of the Geological Society, Vol. LXVI.
    (1910).

Largeness of brain in an individual is no indication of intelligence, and smallness of brain no proof of low mentality. Some of the greatest thinkers, such as Aristotle and Leibnitz, had abnormally small heads. Further, the size of the brain is of no significance whatever except in strict relation to the size and weight of the body. Woman has five or six ounces less brain-matter than man, but in proportion to her average size and the weight of the vital tissue of her body (excluding fat) she has as respectable a brain as man. When, however, these allowances have been made, it has usually been considered that the average brain of a race is in proportion to its average intelligence. This is not strictly true. The rabbit has a larger proportion of brain to body than the elephant or horse, and the canary a larger proportion than the chimpanzee. Professor Sollas says that the average cranial capacity of the Eskimo is 1546 cubic centimetres, or nearly that assigned to the average Parisian.

Clearly the question is very complex, and some of these recent authorities conclude that the cranial capacity, or volume of the brain, has no relation to intelligence, and therefore the size of the Neanderthal skull neither confirms nor disturbs the theory of evolution. The wise man will suspend his judgment until the whole question has been fully reconsidered. But I would point out that some of the recent criticisms are exaggerated. The Gibraltar skull is estimated by Professor Sollas himself to have a capacity of about 1260; and his conclusion that it is an abnormal or feminine skull rests on no positive grounds. The Chapelle-aux-Saints skull ALONE is proved to have the high capacity of 1620; and it is as yet not much more than a supposition that the earlier skulls had been wrongly measured. But, further, the great French authority, M. Boule, who measured the capacity of the Chapelle-aux Saints skull, observes [*] that "the anomaly disappears" on careful study. He assures us that a modern skull of the same dimensions would have a capacity of 1800-1900 cubic centimetres, and warns us that we must take into account the robustness of the body of primitive man. He concludes that the real volume of the Neanderthal brain (in this highest known specimen) is "slight in comparison with the volume of the brain lodged in the large heads of to-day," and that the "bestial or ape-like characters" of the race are not neutralised by this gross measurement.

*See his article in Anthropologie, Vol. XX. (1909), p. 257.
    As Professor Sollas mainly relies on Boule, it is important
        to see that there is a very great difference between the
        two.

We must therefore hesitate to accept the statement that primitive man had as large a brain, if not a larger brain, than a modern race. The basis is slender, and the proportion of brain to body-tissue has not been taken into account. On the other hand, the remains of this early race are, Professor Sollas says, "obviously more brutal than existing men in all the other ascertainable characters by which they differ from them." Nor are we confined to precarious measurements of skulls. We have the remains of the culture of this early race, and in them we have a surer trace of its mental development.

Here again we must proceed with caution, and set aside confused and exaggerated statements. Some refer us to the artistic work of primitive man. We will consider his drawings and carvings presently, but they belong to a later race, not the Neanderthal race. Some lay stress on the fact, apparently indicated in one or two cases out of a dozen, that primitive man buried his dead. Professor Sollas says that it indicates that even Neanderthal man had reached "a comparatively high stage in the evolution of religious ideas "; but the Australians bury their dead, and the highest authorities are not agreed whether they have any idea whatever of a supreme being or of morality. We must also disallow appeals to the use of fire, the taming of animals, pottery, or clothing. None of these things are clearly found in conjunction with the Neanderthal race.

The only certain relic of Neanderthal culture is the implement which the primitive savage fashioned, by chipping or pressure, of flint or other hard stone. The fineness of some of these implements is no indication of great intelligence. The Neanderthal man inherited a stone culture which was already of great antiquity. At least one, if not two or three, prolonged phases of the Old Stone Age were already over when he appeared. On the most modest estimate men had by that time been chipping flints for several hundred thousand years, and it is no argument of general intelligence that some skill in the one industry of the age had been developed. The true measure of Neanderthal man's capacity is that, a million years or so after passing the anthropoid-age level, he chipped his stones more finely and gave them a better edge and contour. There is no evidence that he as yet hefted them. It is flattering to him to compare him with the Australian aboriginal. The native art, the shields and spears and boomerangs, and the elaborate tribal and matrimonial arrangements of the Australian black are not known to have had any counterpart in his life.

It would therefore seem that the precursors of man made singularly little, if any, progress during the vast span of time between the Miocene and the Ice-Age, and that then something occurred which quickened the face of human evolution. From the Neanderthal level man will advance to the height of modern civilisation in about one-tenth the time that it took him to advance from the level of the higher ape to that of the lowest savage. Something has broken into the long lethargy of his primitive career, and set him upon a progressive path. Let us see if a careful review of the stages of his culture confirms the natural supposition that this "something" was the fall in the earth's temperature, and how it may have affected him.

CHAPTER XX. THE DAWN OF CIVILISATION

The story of man before the discovery of metal and the attainment of civilisation is notoriously divided into a Palaeolithic (Old Stone) Age, and a Neolithic (New Stone) Age. Each of these ages is now subdivided into stages, which we will review in succession. But it is important to conceive the whole story of man in more correct proportion than this familiar division suggests. The historical or civilised period is now computed at about ten thousand years. The Neolithic Age, which preceded civilisation, is usually believed to be about four or five times as long, though estimates of its duration vary from about twenty to a hundred thousand years. The Palaeolithic Age in turn is regarded as at least three or four times as long as the Neolithic; estimates of time vary from a hundred to five hundred thousand years. And before this there is the vast stretch of time in which the ape slowly became a primitive human.

This long, early period is, as we saw, still wrapped in mist and controversy. A few bones tell of a race living, in semi-human shape, in the region of the Indian Ocean; a few crude stones are held by many to indicate that a more advanced, but very lowly race, wandered over the south of Europe and north of Africa before the Ice-Age set in. The starting-point or cradle of the race is not known. The old idea of seeking the patriarchal home on the plains to the north of India is abandoned, and there is some tendency to locate it in the land which has partly survived in the islands of the Indian Ocean. The finding of early remains in Java is not enough to justify that conclusion, but it obtains a certain probability when we notice the geographical distribution of the Primates. The femurs and the apes are found to-day in Africa and Asia alone; the monkeys have spread eastward to America and westward to Europe and Africa; the human race has spread north-eastward into Asia and America, northwestward into Europe, westward into Africa, and southward to Australia and the islands. This distribution suggests a centre in the Indian Ocean, where there was much more land in the Tertiary Era than there is now. We await further exploration in that region and Africa.

There is nothing improbable in the supposition that man wandered into Europe in the Tertiary, and has left in the Eoliths the memorials of his lowly condition. The anthropoid apes certainly reached France. However that may be, the Ice-Age would restrict all the Primates to the south. It will be seen, on a glance at the map, that a line of ice-clad mountains would set a stern barrier to man's advance in the early Pleistocene, from the Pyrenees to the Himalaya, if not to the Pacific. He therefore spread westward and southward. One branch wandered into Australia, and was afterwards pressed by more advanced invaders (the present blacks of Australia) into Tasmania, which seems to have been still connected by land. Another branch, or branches, spread into Africa, to be driven southward, or into the central forests, by later and better equipped invaders. They survive, little changed (except by recent contact with Europeans), in the Bushmen and in large populations of Central Africa which are below the level of tribal organisation. Others remained in the islands, and we seem to have remnants of them in the Kalangs, Veddahs, etc. But these islands have been repeatedly overrun by higher races, and the primitive life has been modified.

Comparing the most isolated of these relics of early humanity, we obtain many suggestions about the life of that remote age. The aboriginal Tasmanians, who died out about forty years ago, were of great evolutionary interest. It is sometimes said that man is distinguished from all other animals by the possession of abstract ideas, but the very imperfect speech of the Tasmanians expressed no abstract ideas. Their mind seems to have been in an intermediate stage of development. They never made fire, and, like the other surviving fragments of early humanity, they had no tribal organisation, and no ideas of religion or morality.

The first effect of the Ice-Age on this primitive humanity would be to lead to a beginning of the development of racial characters. The pigment under the skin of the negro is a protection against the actinic rays of the tropical sun; the white man, with his fair hair and eyes, is a bleached product of the northern regions; and the yellow or brown skin seems to be the outcome of living in dry regions with great extremes of temperature. As the northern hemisphere divided into climatic zones these physical characters were bound to develop. The men who went southward developed, especially when fully exposed to the sun on open plains, the layer of black pigment which marks the negroid type. There is good reason, as we shall see to think that man did not yet wear clothing, though he had a fairly conspicuous, if dwindling, coat of hair. On the other hand the men who lingered further north, in South-western Asia and North Africa, would lose what pigment they had, and develop the lighter characters of the northerner. It has been noticed that even a year in the arctic circle has a tendency to make the eyes of explorers light blue. We may look for the genesis of the vigorous, light-complexioned races along the fringe of the great ice-sheet. It must be remembered that when the limit of the ice-sheet was in Central Germany and Belgium, the climate even of North Africa would be very much more temperate than it is to-day.

As the ice-sheet melted, the men who were adapted to living in the temperate zone to the south of it penetrated into Europe, and the long story of the Old Stone Age opened. It must not, of course, be supposed that this stage of human culture only began with the invasion of Europe. Men would bring their rough art of fashioning implements with them, but the southern regions are too little explored to inform us of the earlier stage. But as man enters Europe he begins to drop his flints on a soil that we have constant occasion to probe—although the floor on which he trod is now sometimes forty or fifty feet below the surface—and we obtain a surer glimpse of the fortunes of our race.

Most European geologists count four distinct extensions of the ice-sheet, with three interglacial periods. It is now generally believed that man came north in the third interglacial period; though some high authorities think that he came in the second. As far as

England is concerned, it has been determined, under the auspices of the British Association, that our oldest implements (apart from the Eoliths) are later than the great ice-sheet, but there is some evidence that they precede the last extension of the ice.

Two stages are distinguished in this first part of the Palaeolithic Age—the Acheulean and Chellean—but it will suffice for our purpose to take the two together as the earlier and longer section of the Old Stone Age. It was a time of temperate, if not genial, climate. The elephant (an extinct type), the rhinoceros, the hippopotamus, the hyaena, and many other forms of animal life that have since retired southward, were neighbours of the first human inhabitant of Europe. Unfortunately, we have only one bone of this primitive race, the jaw found at Mauer in 1907, but its massive size and chinless contour suggest a being midway between the Java man and the Neanderthal race. His culture confirms the supposition. There is at this stage no clear trace of fire, clothing, arrows, hefted weapons, spears, or social life. As the implements are generally found on old river-banks or the open soil, not in caves, we seem to see a squat and powerful race wandering, homeless and unclad, by the streams and broad, marshy rivers of the time. The Thames and the Seine had not yet scooped out the valleys on the slopes of which London and Paris are built.

This period seems, from the vast number of stone implements referred to it, to have lasted a considerable time. There is a risk in venturing to give figures, but it may be said that few authorities would estimate it at less than a hundred thousand years. Man still advanced with very slow and uncertain steps, his whole progress in that vast period being measured by the invention of one or two new forms of stone implements and a little more skill in chipping them. At its close a great chill comes over Europe—the last ice-sheet is, it seems, spreading southward—and we enter the Mousterian period and encounter the Neanderthal race which we described in the preceding chapter.

It must be borne in mind that the whole culture of primitive times is crushed into a few feet of earth. The anthropologist is therefore quite unable to show us the real succession of human stages, and has to be content with a division of the whole long and gradual evolution into a few well-marked phases. These phases, however, shade into each other, and are merely convenient measurements of a continuous story. The Chellean man has slowly advanced to a high level. There is no sudden incoming of a higher culture or higher type of man. The most impressive relics of the Mousterian period, which represent its later epoch, are merely finely chipped implements. There is no art as yet, no pottery, and no agriculture; and there is no clear trace of the use of fire or clothing, though we should be disposed to put these inventions in the chilly and damp Mousterian period. There is therefore no ground for resenting the description, "the primeval savage," which has been applied to early man. The human race is already old, yet, as we saw, it is hardly up to the level of the Australian black. The skeleton found at Chapelle-aux-Saints is regarded as the highest known type of the race, yet the greatest authority on it, M. Boule, says emphatically: "In no actual race do we find the characters of inferiority—that is to say, the ape-like features—which we find in the Chapelle-aux-Saints head." The largeness of the head is in proportion to the robust frame, but in its specifically human part—the front—it is very low and bestial; while the heavy ridges over the large eyes, the large flat stumpy nose, the thick bulge of the lips and teeth, and the almost chinless jaw, show that the traces of his ancestry cling close to man after some hundreds of thousands of years of development.

The cold increases as we pass to the last part of the Old Stone Age, the Solutrean and Magdalenian periods; and nothing is clearer than that the pace of development increases at the same time. Short as the period is, in comparison with the preceding, it witnesses a far greater advance than had been made in all the rest of the Old Stone Age. Beyond a doubt men now live in caves, in large social groups, make clothing from the skins of animals, have the use of fire, and greatly improve the quality of their stone axes, scrapers, knives, and lance-heads. There is at last some promise of the civilisation that is coming. In the soil of the caverns in which man lived, especially in Southern France and the Pyrenean region, we find the debris of a much larger and fuller life. Even the fine bone needles with which primitive man sewed his skin garments, probably with sinews for thread, survive in scores. In other places we find the ashes of the fires round which he squatted, often associated with the bones of the wild horses, deer, etc., on which he lived.

But the most remarkable indication of progress in the "cave-man" is his artistic skill. Exaggerated conclusions are sometimes drawn from the statuettes, carvings, and drawings which we find among the remains of Magdalenian life. Most of them are crude, and have the limitations of a rustic or a child artist. There is no perspective, no grouping. Animals are jumbled together, and often left unfinished because the available space was not measured. There are, however, some drawings—cut on bone or horn or stone with a flint implement—which evince great skill in line-drawing and, in a few cases, in composition. Some of the caves also are more or less frescoed; the outlines of animals, sometimes of life-size and in great numbers, are cut in the wall, and often filled in with pigment. This skill does not imply any greater general intelligence than the rest of the culture exhibits. It implies persistent and traditional concentration upon the new artistic life. The men who drew the "reindeer of Thayngen" and carved the remarkable statuettes of women in ivory or stone, were ignorant of the simplest rudiments of pottery or agriculture, which many savage tribes possess.

Some writers compare them with the Eskimo of to-day, and even suggest that the Eskimo are the survivors of the race, retreating northward with the last ice-sheet, and possibly egged onward by a superior race from the south. It is, perhaps, not a very extravagant claim that some hundreds of thousands of years of development—we are now only a few tens of thousands of years from the dawn of civilisation—had lifted man to the level of the Eskimo, yet one must hesitate to admit the comparison. Lord Avebury reproduces an Eskimo drawing, or picture-message, in his "Prehistoric Times," to which it would be difficult to find a parallel in Magdalenian remains. I do not mean that the art is superior, but the complex life represented on the picture-message, and the intelligence with which it is represented, are beyond anything that we know of Palaeolithic man. I may add that nearly all the drawings and statues of men and women which the Palaeolithic artist has left us are marked by the intense sexual exaggeration—the "obscenity," in modern phraseology—which we are apt to find in coarse savages.

Three races are traced in this period. One, identified by skeletons found at Mentone and by certain statuettes, was negroid in character. Probably there was an occasional immigration from Africa. Another race (Cro-Magnon) was very tall, and seems to represent an invasion from some other part of the earth toward the close of the Old Stone Age. The third race, which is compared to the Eskimo, and had a stature of about five feet, seem to be the real continuers of the Palaeolithic man of Europe. Curiously enough, we have less authentic remains of this race than of its predecessor, and can only say that, as we should expect, the ape-like features—the low forehead, the heavy frontal ridges, the bulging teeth, etc.—are moderating. The needles we have found—round, polished, and pierced splinters of bone, sometimes nearly as fine as a bodkin—show indisputably that man then had clothing, but it is curious that the artist nearly always draws him nude. There is also generally a series of marks round the contour of the body to indicate that he had a conspicuous coat of hair. Unfortunately, the faces of the men are merely a few unsatisfactory gashes in the bone or horn, and do not picture this interesting race to us. The various statuettes of women generally suggest a type akin to the wife of the Bushman.

We have, in fine, a race of hunters, with fine stone knives and javelins. Toward the close of the period we find a single representation of an arrow, which was probably just coming into use, but it is not generally known in the Old Stone Age. One of the drawings seems to represent a kind of bridle on a horse, but we need more evidence than this to convince us that the horse was already tamed, nor is there any reason to suppose that the dog or reindeer had been tamed, or that the ground was tilled even in the most rudimentary way. Artistic skill, the use of clothing and fire, and a finer feeling in the shaping of weapons and implements, are the highest certain indications of the progress made by the end of the Old Stone Age.

But there was probably an advance made which we do not find recorded, or only equivocally recorded, in the memorials of the age. Speech was probably the greatest invention of Magdalenian man. It has been pointed out that the spine in the lower jaw, to which the tongue-muscle is attached, is so poorly developed in Palaeolithic man that we may infer from it the absence of articulate speech. The deduction has been criticised, but a comparison of the Palaeolithic jaw with that of the ape on one hand and modern man on the other gives weight to it. Whatever may have been earlier man's power of expression, the closer social life of the Magdalenian period would lead to a great development of it. Some writers go so far as to suggest that certain obscure marks painted on pebbles or drawn on the cavern-walls by men at the close of the Palaeolithic Age may represent a beginning of written language, or numbers, or conventional signs. The interpretation of these is obscure and doubtful. It is not until ages afterwards that we find the first clear traces of written language, and then they take the form of pictographs (like the Egyptian hieroglyphics or the earliest Chinese characters).

We cannot doubt, however, that articulate speech would be rapidly evolved in the social life of the later Magdalenian period, and the importance of this acquisition can hardly be exaggerated. Imagine even a modern community without the device of articulate language. A very large proportion of the community, who are now maintained at a certain level by the thought of others, communicated to them by speech, would sink below the civilised standard, and the transmission and improvement of ideas would be paralysed. It would not be paradoxical to regard the social life and developing speech of Magdalenian man as the chief cause of the rapid advance toward civilisation which will follow in the next period.

And it is not without interest to notice that a fall in the temperature of the earth is the immediate cause of this social life. The building of homes of any kind seems to be unknown to Magdalenian man. The artist would have left us some sketchy representation of it if there had been anything in the nature of a tent in his surroundings. The rock-shelter and the cave are the homes which men seek from the advancing cold. As these are relatively few in number, fixed in locality, and often of large dimensions, the individualism of the earlier times is replaced by collective life. Sociologists still dispute whether the clan arose by the cohesion of families or the family arose within the clan. Such evidence as is afforded by prehistoric remains is entirely in favour of the opinion of Professor Westermarck, that the family preceded the larger group. Families of common descent would now cling together and occupy a common cavern, and, when the men gathered at night with the women for the roasting and eating of the horse or deer they had hunted, and the work of the artist and the woman was considered, the uncouth muttering and gesticulating was slowly forged into the great instrument of articulate speech. The first condition of more rapid progress was instinctively gained.

Our story of life has so often turned on this periodical lowering of the climate of the earth that it is interesting to find this last and most important advance so closely associated with it that we are forced once more to regard it as the effective cause. The same may be said of another fundamental advance of the men of the later Palaeolithic age, the discovery of the art of making fire. It coincides with the oncoming of the cold, either in the Mousterian or the Magdalenian. It was more probably a chance discovery than an invention. Savages so commonly make fire by friction—rubbing sticks, drills, etc.—that one is naturally tempted to regard this as the primitive method. I doubt if this was the case. When, in Neolithic times, men commonly bury the dead, and put some of their personal property in the grave with them, the fire-kindling apparatus we find is a flint and a piece of iron pyrites. Palaeolithic man made his implements of any kind of hard and heavy stone, and it is probable that he occasionally selected iron ore for the purpose. An attempt to chip it with flint would cause sparks that might fall on inflammable material, and set it alight. Little intelligence would be needed to turn this discovery to account.

Apart from these conjectures as to particular features in the life of prehistoric man, it will be seen that we have now a broad and firm conception of its evolution. From the ape-level man very slowly mounts to the stage of human savagery. During long ages he seems to have made almost no progress. There is nothing intrinsically progressive in his nature. Let a group of men be isolated at any stage of human evolution, and placed in an unchanging environment, and they will remain stationary for an indefinite period. When Europeans began to traverse the globe in the last few centuries, they picked up here and there little groups of men who had, in their isolation, remained just where their fathers had been when they quitted the main road of advance in the earlier stages of the Old Stone Age. The evolution of man is guided by the same laws as the evolution of any other species. Thus we can understand the long period of stagnation, or of incalculably slow advance. Thus, too, we can understand why, at length, the pace of man toward his unconscious goal is quickened. He is an inhabitant of the northern hemisphere, and the northern hemisphere is shaken by the last of the great geological revolutions. From its first stress emerges the primeval savage of the early part of the Old Stone Age, still bearing the deep imprint of his origin, surpassing his fellow-animals only in the use of crude stone implements. Then the stress of conditions relaxes—the great ice-sheet disappears—and again during a vast period he makes very little progress. The stress returns. The genial country is stripped and impoverished, and the reindeer and mammoth spread to the south of Europe. But once more the adversity has its use, and man, stimulated in his hunt for food, invigorated by the cold, driven into social life, advances to the culmination of the Old Stone Age.

We are still very far from civilisation, but the few tens of thousands of years that separate Magdalenian man from it will be traversed with relative speed—though, we should always remember, with a speed far less than the pace at which man is advancing to-day. A new principle now enters into play: a specifically human law of evolution is formulated. It has no element of mysticism, and is merely an expression of the fact that the previous general agencies of development have created in man an intelligence of a higher grade than that of any other animal. In his larger and more plastic brain the impressions received from the outer world are blended in ideas, and in his articulate speech he has a unique means of entering the idea-world of his fellows. The new principle of evolution, which arises from this superiority, is that man's chief stimulus to advance will now come from his cultural rather than his physical environment. Physical surroundings will continue to affect him. One race will outstrip another because of its advantage in soil, climate, or geographical position. But the chief key to the remaining and more important progress of mankind, which we are about to review, is the stimulating contact of the differing cultures of different races.

This will be seen best in the history of civilisation, but the principle may be recognised in the New Stone Age which leads from primeval savagery to civilisation, or, to be more accurate and just, to the beginning of the historical period. It used to be thought that there was a mysterious blank or gulf between the Old and the New Stone Age. The Palaeolithic culture seemed to come to an abrupt close, and the Neolithic culture was sharply distinguished from it. It was suspected that some great catastrophe had destroyed the Palaeolithic race in Europe, and a new race entered as the adverse conditions were removed. This was especially held to be the case in England. The old Palaeolithic race had never reached Ireland, which seems to have been cut oft from the Continent during the Ice-Age, and most of the authorities still believe— in spite of some recent claims—that it never reached Scotland. England itself was well populated, and the remains found in the caves of Derbyshire show that even the artist—or his art—had reached that district. This Palaeolithic race seemed to come to a mysterious end, and Europe was then invaded by the higher Neolithic race. England was probably detached from the Continent about the end of the Magdalenian period. It was thought that some great devastation—the last ice-sheet, a submersion of the land, or a plague—then set in, and men were unable to retreat south.

It is now claimed by many authorities that there are traces of a Middle Stone (Mesolithic) period even in England, and nearly all the authorities admit that such a transitional stage can be identified in the Pyrenean region. This region had been the great centre of the Magdalenian culture. Its large frescoed caverns exhibit the culmination of the Old Stone life, and afford many connecting links with the new. It is, however, a clearly established and outstanding fact that the characteristic art of Magdalenian man comes to an abrupt and complete close, and it does not seem possible to explain this without supposing that the old race was destroyed or displaced. If we could accept the view that it was the Eskimo-like race of the Palaeolithic that cultivated this art, and that they retreated north with the reindeer and the ice, and survive in our Eskimo, we should have a plausible explanation. In point of fact, we find no trace whatever of this slow migration from the south of Europe to the north. The more probable supposition is that a new race, with more finished stone implements, entered Europe, imposed its culture upon the older race, and gradually exterminated or replaced it. We may leave it open whether a part of the old race retreated to the north, and became the Eskimo.

Whence came the new race and its culture? It will be seen on reflection that we have so far been studying the evolution of man in Europe only, because there alone are his remains known with any fullness. But the important region which stretches from Morocco to Persia must have been an equally, if not more, important theatre of development. While Europe was shivering in the last stage of the Ice-Age, and the mammoth and reindeer browsed in the snows down to the south of France, this region would enjoy an excellent climate and a productive soil. We may confidently assume that there was a large and stirring population of human beings on it during the Magdalenian cold. We may, with many of the authorities, look to this temperate and fertile region for the slight advance made by early Neolithic man beyond his predecessor. As the cold relaxed, and the southern fringe of dreary steppe w as converted once more into genial country, the race would push north. There is evidence that there were still land bridges across the Mediterranean. From Spain and the south of France this early Neolithic race rapidly spread over Europe.

It must not be supposed that the New Stone Age at first goes much beyond the Old in culture. Works on prehistoric man are apt to give as features of "Neolithic man" all that we know him to have done or discovered during the whole of the New Stone Age. We read that he not only gave a finer finish to, and sometimes polished, his stone weapons, but built houses, put imposing monuments over his dead, and had agriculture, tame cattle, pottery, and weaving. This is misleading, as the more advanced of these accomplishments appear only late in the New Stone Age. The only difference we find at first is that the stone axes, etc., are more finely chipped or flaked, and are frequently polished by rubbing on stone moulds.

There is no sudden leap in culture or intelligence in the story of man.

It would be supremely interesting to trace the evolution of human industries and ideas during the few tens of thousands of years of the New Stone Age. During that time moral and religious ideas are largely developed, political or social forms are elaborated, and the arts of civilised man have their first rude inauguration. The foundations of civilisation are laid. Unfortunately, precisely because the period is relatively so short and the advance so rapid, its remains are crushed and mingled in a thin seam of the geological chronicle, and we cannot restore the gradual course of its development with any confidence. Estimates of its duration vary from 20,000 to 70,000 years; though Sir W. Turner has recently concluded, from an examination of marks on Scottish monuments, that Neolithic man probably came on foot from Scandinavia to Scotland, and most geologists would admit that it must be at least a hundred thousand years since one could cross from Norway to Scotland on foot. As usual, we must leave open the question of chronology, and be content with a modest provisional estimate of 40,000 or 50,000 years.

We dimly perceive the gradual advance of human culture in this important period. During the Old Stone Age man had made more progress than he had made in the preceding million years; during the New Stone Age—at least one-fourth as long as the Old—he made even greater progress; and, we may add, in the historical period, which is one-fourth the length of the Neolithic Age, he will make greater progress still. The pace of advance naturally increases as intelligence grows, but that is not the whole explanation. The spread of the race, the gathering of its members into tribes, and the increasing enterprise of men in hunting and migration, lead to incessant contacts of different cultures and a progressive stimulation.

At first Neolithic man is content with finer weapons. His stone axe is so finely shaped and polished that it sometimes looks like forged or moulded metal. He also drills a clean hole through it—possibly by means of a stick working in wet sand—and gives it a long wooden handle. He digs in the earth for finer flints, and in some of his ancient shafts (Grimes, Graves and Cissbury) we find picks of reindeer horn and hollowed blocks of chalk in which he probably burned fat for illumination underground. But in the later part of the Neolithic—to which much of this finer work also may belong—we find him building huts, rearing large stone monuments, having tame dogs and pigs and oxen, growing corn and barley, and weaving primitive fabrics. He lives in large and strong villages, round which we must imagine his primitive cornfields growing and his cattle grazing, and in which there must have been some political organisation under chiefs.

When we wish to trace the beginning of these inventions we have the same difficulty that we experienced in tracing the first stages of new animal types. The beginning takes place in some restricted region, and our casual scratching of the crust of the earth or the soil may not touch it for ages, if it has survived at all. But for our literature and illustrations a future generation would be equally puzzled to know how we got the idea of the aeroplane or the electric light. In some cases we can make a good guess at the origin of Neolithic man's institutions. Let us take pottery. Palaeolithic man cooked his joint of horse or reindeer, and, no doubt, scorched it. Suppose that some Palaeolithic Soyer had conceived the idea of protecting the joint, and preserving its juices, by daubing it with a coat of clay. He would accidentally make a clay vessel. This is Mr. Clodd's ingenious theory of the origin of pottery. The development of agriculture is not very puzzling. The seed of corn would easily be discovered to have a food-value, and the discovery of the growth of the plant from the seed would not require a very high intelligence. Some ants, we may recall, have their fungus-beds. It would be added by many that the ant gives us another parallel in its keeping of droves of aphides, which it "milks." But it is now doubted if the ant deliberately cultivates the aphides with this aim. Early weaving might arise from the plaiting of grasses. If wild flax were used, it might be noticed that part of it remained strong when the rest decayed, and so the threads might be selected and woven.

The building of houses, after living for ages in stone caverns, would not be a very profound invention. The early houses were—as may be gathered from the many remains in Devonshire and Cornwall—mere rings of heaped stones, over which, most probably, was put a roof of branches or reeds, plastered with mud. They belong to the last part of the New Stone Age. In other places, chiefly Switzerland, Neolithic man lived in wooden huts built on piles in the shallow shores of lakes. It is an evidence that life on land is becoming as stimulating as we find it in the age of Deinosaurs or early mammals. These pile-villages of Switzerland lasted until the historical period, and the numerous remains in the mud of the lake show the gradual passage into the age of metal.

Before the metal age opened, however, there seem to have been fresh invasions of Europe and changes of its culture. The movements of the various early races of men are very obscure, and it would be useless to give here even an outline of the controversy. Anthropologists have generally taken the relative length and width of the skull as a standard feature of a race, and distinguished long-headed (dolichocephalic), short-headed (brachycephalic), and middle-headed (mesaticephalic) races. Even on this test the most divergent conclusions were reached in regard to early races, and now the test itself is seriously disputed. Some authorities believe that there is no unchanging type of skull in a particular race, but that, for instance, a long-headed race may become short-headed by going to live in an elevated region.

It may be said, in a few words, that it is generally believed that two races invaded Europe and displaced the first Neolithic race. The race which chiefly settled in the Swiss region is generally believed to have come from Asia, and advanced across Europe by way of the valley of the Danube. The native home of the wheat and barley and millet, which, as we know, the lake-dwellers cultivated, is said to be Asia. On the other hand, the Neolithic men who have left stone monuments on our soil are said to be a different race, coming, by way of North Africa, from Asia, and advancing along the west of Europe to Scandinavia. A map of the earth, on which the distribution of these stone monuments—all probably connected with the burial of the dead—is indicated, suggests such a line of advance from India, with a slighter branch eastward. But the whole question of these invasions is disputed, and there are many who regard the various branches of the population of Europe as sections of one race which spread upward from the shores of the Mediterranean.

It is clear at least that there were great movements of population, much mingling of types and commercial interchange of products, so that we have the constant conditions of advance. A last invasion seems to have taken place some two or three thousand years before the Christian era, when the Aryans overspread Europe. After all the controversy about the Aryans it seems clear that a powerful race, representing the ancestors of most of the actual peoples of Europe and speaking the dialects which have been modified into the related languages of the Greeks, Romans, Germans, Celts, Lithuanians, etc., imposed its speech on nearly the whole of the continent. Only in the Basques and Picts do we seem to find some remnants of the earlier non-Aryan tongues. But whether these Aryans really came from Asia, as it used to be thought, or developed in the east of Europe, is uncertain. We seem justified in thinking that a very robust race had been growing in numbers and power during the Neolithic Age, somewhere in the region of South-east Europe and Southwest Asia, and that a few thousand years before the Christian Era one branch of it descended upon India, another upon the Persian region, and another overspread Europe. We will return to the point later. Instead of being the bearers of a higher civilisation, these primitive Aryans seem to have been lower in culture than the peoples on whom they fell.

The Neolithic Age had meantime passed into the Age of Metal. Copper was probably the first metal to be used. It is easily worked, and is found in nature. But the few copper implements we possess do not suggest a "Copper Age" of any length or extent. It was soon found, apparently, that an admixture of tin hardened the copper, and the Bronze Age followed. The use of bronze was known in Egypt about 4800 B.C. (Flinders Petrie), but little used until about 2000 B.C. By that time (or a few centuries later) it had spread as far as Scandinavia and Britain. The region of invention is not known, but we have large numbers of beautiful specimens of bronze work—including brooches and hair-pins—in all parts of Europe. Finally, about the thirteenth century B.C., we find the first traces of the use of iron. The first great centre for the making of iron weapons seems to have been Hallstatt, in the Austrian Alps, whence it spread slowly over Europe, reaching Scandinavia and Britain between 500 and 300 B.C. But the story of man had long before this entered the historical period, to which we now turn.

CHAPTER XXI. EVOLUTION IN HISTORY

In the preceding chapters I have endeavoured to show how, without invoking any "definitely directed variations," which we seem to have little chance of understanding, we may obtain a broad conception of the way in which the earth and its living inhabitants came to be what they are. No one is more conscious than the writer that this account is extremely imperfect. The limits of the volume have permitted me to use only a part of the material which modern science affords, but if the whole of our discoveries were described the sketch would still remain very imperfect. The evolutionary conception of the world is itself undergoing evolution in the mind of man. Age by age the bits of fresh discovery are fitted into the great mosaic. Large areas are still left for the scientific artist of the future to fill. Yet even in its imperfect state the evolutionary picture of the world is most illuminating. The questions that have been on the lips of thoughtful men since they first looked out with adult eyes on the panorama of nature are partly answered. Whence and Why are no longer sheer riddles of the sphinx.

It remains to be seen if evolutionary principles will throw at least an equal light on the progress of humanity in the historical period. Here again the questions, Whence and Why, have been asked in vain for countless ages. If man is a progressive animal, why has the progress been confined to some of the race? If humanity shared at first a common patrimony, why have the savages remained savages, and the barbarians barbaric? Why has progress been incarnated so exceptionally in the white section of the race, the Europeans? We approach these questions more confidently after surveying the story of terrestrial life in the light of evolutionary principles. Since the days of the primeval microbe it has happened that a few were chosen and many were left behind. There was no progressive element in the advancing few that was not shared by the stagnant many. The difference lay in the environment. Let us see if this principle applies to the history of civilisation.

In the last chapter I observed that, with the rise of human intelligence, the cultural environment becomes more important than the physical. Since human progress is a progress in ideas and the emotions which accompany them, this may seem to be a truism. In point of fact it is assailed by more than one recent historical writer. The scepticism is partly due to a misunderstanding. No one but a fanatical adherent of extreme theories of heredity will deny that the physical surroundings of a race continue to be of great importance. The progress of a particular people may often be traced in part to its physical environment; especially to changes of environment, by migration, for instance. Further, it is not for a moment suggested that a race never evolves its own culture, but has always to receive it from another. If we said that, we should be ultimately driven to recognise culture, like the early Chinese, as a gift of the gods. What is meant is that the chief key to the progress of certain peoples, the arrest of progress in others, and the entire absence of progress in others, is the study of their relations with, or isolation from, other peoples. They make progress chiefly according to the amount of stimulation they get by contact with a diverse culture.

Let us see if this furnishes a broad explanation of the position of the various peoples of the world. The Ethnologist tells us that the lowest peoples of the earth are the Yahgans of Tierra del Fuego, the Hottentots, a number of little-understood peoples in Central Africa, the wild Veddahs of Ceylon, the (extinct) Tasmanians, the Aetas in the interior of the Philippines, and certain fragments of peoples on islands of the Indian Ocean. There is not the least trace of a common element in the environment of these peoples to explain why they have remained at the level of primitive humanity. Many of them lived in the most promising and resourceful surroundings. What is common to them all is their isolation from the paths of later humanity. They represent the first wave of human distribution, pressed to the tips of continents or on islands by later waves, and isolated. The position of the Veddahs is, to some extent, an exception; and it is interesting to find that the latest German students of that curious people think that they have been classed too low by earlier investigators.

We cannot run over all the peoples of the earth in this way, but will briefly glance at the lower races of the various continents. A branch of the second phase of developing humanity, the negroid stock, spread eastward over the Asiatic islands and Australia, and westward into Africa. The extreme wing of this army, the Australian blacks, too clearly illustrates the principle to need further reference. It has retained for ages the culture of the middle Palaeolithic. The negritos who penetrated to the Philippines are another extreme instance of isolation. The Melanesians of the islands of the Indian and Pacific Ocean are less low, because those islands have been slowly crossed by a much higher race, the Polynesians. The Maoris of New Zealand, the Tongans, Hawaians, etc., are people of our own (Caucasic) stock, probably diverging to the south-east while our branch of the stock pressed westward. This not only explains the higher condition of the Maoris, etc., but also shows why they have not advanced like their European cousins. Their environment is one of the finest in the world, but—it lies far away from the highways of culture.

In much the same way can we interpret the swarming peoples of Africa. The more primitive peoples which arrived first, and were driven south or into the central forests by the later and better equipped invaders from the central zone, have remained the more primitive. The more northern peoples, on the fringe of, or liable to invasion from, the central zone, have made more advance, and have occasionally set up rudimentary civilisations. But the movements from the north to the south in early historical times are too obscure to enable us to trace the action of the principle more clearly. The peoples of the Mediterranean fringe of Africa, living in the central zone of stimulation, have proved very progressive. Under the Romans North Africa was at least as civilised as Britain, and an equally wise and humane European policy might lead to their revival to-day.

When we turn to Asia we encounter a mass of little-understood peoples and a few civilisations with obscure histories, but we have a fairly clear application of the principle. The northern, more isolated peoples, are the more primitive; the north-eastern, whose isolation is accentuated by a severe environment, are most primitive of all. The Eskimo, whether they are the survivors of the Magdalenian race or a regiment thrown off the Asiatic army as it entered America, remain at the primitive level. The American peoples in turn accord with this view. Those which penetrate furthest south remain stagnant or deteriorate; those which remain in the far north remain below the level of civilisation, because the land-bridge to Asia breaks down; but those which settle in Central America evolve a civilisation. A large zone, from Mexico to Peru, was overspread by this civilisation, and it was advancing steadily when European invaders destroyed it, and reduced the civilised Peruvians to the Quichas of to-day.

There remain the civilisations of Asia, and here we have a new and interesting aspect of the question. How did these civilisations develop in Asia, and how is it that they have remained stagnant for ages, while Europe advanced? The origin of the Asiatic civilisations is obscure. The common idea of their vast antiquity has no serious ground. The civilisation of Japan cannot be traced back beyond about the eighth century B.C. Even then the population was probably a mixed flotsam from neighbouring lands—Ainus, Koreans, Chinese, and Malays. What was the character of the primitive civilisation resulting from the mixture

366

of these different cultures we do not know. But the chief elements of Japanese civilisation came later from China. Japan had no written language of any kind until it received one from China about the sixth century of the Christian Era.

The civilisation of China itself goes back at least to about 2300 B.C., but we cannot carry it further back with any confidence. The authorities, endeavouring to pick their steps carefully among old Chinese legends, are now generally agreed that the primitive Chinese were a nomadic tribe which slowly wandered across Asia from about the shores of the Caspian Sea. In other words, they started from a region close to the cradle of western civilisation. Some students, in fact, make them akin to the Akkadians, who founded civilisation in Mesopotamia. At all events, they seem to have conveyed a higher culture to the isolated inhabitants of Western Asia, and a long era of progress followed their settlement in a new environment. For more than two thousand years, however, they have been enclosed in their walls and mountains and seas, while the nations of the remote west clashed unceasingly against each other. We need no other explanation of their stagnation. To speak of the "unprogressiveness" of the Chinese is pure mysticism. The next generation will see.

The civilisation of India is also far later than the civilisation of the west, and seems to be more clearly due to borrowing from the west. The primitive peoples who live on the hills about India, or in the jungles, are fragments, apparently, of the Stone Age inhabitants of India, or their descendants. Their culture may have degenerated under the adverse conditions of dislodgement from their home, but we may fairly conclude that it was never high. On these primitive inhabitants of the plains of India there fell, somewhere about or before 1000 B.C., the Asiatic branch of the Aryan race.

A very recent discovery (1908) has strongly confirmed and illumined this view of the origin of Indian civilisation. Explorers in the ruins of the ancient capital of the Hittite Empire (in North Syria and Cappadocia) found certain treaties which had been concluded, about 1300 B.C., between the Hittites and the king of the Aryans. The names of the deities which are mentioned in the treaties seem to show that the Persian and Indian branches of the Aryan race were not yet separated, but formed a united kingdom on the banks of the Euphrates. They seem to have come from Bactria (and possibly beyond), and introduced the horse (hitherto unknown to the Babylonians) about 1800 B.C. It is surmised by the experts that the Indian and Persian branches separated soon after 1300 B.C., possibly on account of religious quarrels, and the Sanscrit-speaking branch, with its Vedic hymns and its Hinduism, wandered eastward and northward until it discovered and took possession of the Indian peninsula. The long isolation of India, since the cessation of its commerce with Rome until modern times, explains the later stagnation of its civilisation.

Thus the supposed "non-progressiveness" of the east, after once establishing civilisation, turns out to be a question of geography and history. We have now to see if the same intelligible principles will throw light on the "progressiveness" of the western branch of the Aryan race, and on the course of western civilisation generally. [*]

The first two centres of civilisation are found in the valley of the Nile and the valley of the Tigris and Euphrates; the civilisations of Egypt and Babylon, the oldest in the world. There is, however, a good deal of evidence by which we may bring these civilisations nearer to each other in their earliest stages, so that we must not confidently speak of two quite independent civilisations. The civilisation which developed on the Euphrates is found first at Susa, on the hills overlooking the plains of Mesopotamia, about 6000 B.C. A people akin to the Turkish or Chinese lives among the hills, and makes the vague advance from higher Neolithic culture to primitive civilisation. About the same time the historical or dynastic civilisation begins in Egypt, and some high authorities, such as Mr. Flinders Petrie, believe that the evidence suggests that the founders of this dynastic civilisation came from "the mountainous region between Egypt and the Red Sea." From the northern part of the same region, we saw, the ancestors of the Chinese set out across Asia.

We have here a very suggestive set of facts in connection with early civilisation. The Syro-Arabian region seems to have been a thickly populated centre of advancing tribes, which would be in striking accord

369

with the view of progress that I am following. But we need not press the disputed and obscure theory of the origin of the historic Egyptians. The remains are said to show that the lower valley of the Nile, which must have been but recently formed by the river's annual deposit of mud, was a theatre of contending tribes from about 8000 to 6000 B.C. The fertile lands that had thus been provided attracted tribes from east, west, and south, and there is a great confusion of primitive cultures on its soil.

It is not certain that the race which eventually conquered and founded the historical dynasties came from the mountainous lands to the east. It is enough for us to know that the whole region fermented with jostling peoples. Why it did so the previous chapters will explain. It is the temperate zone into which men had been pressed by the northern ice-sheet, and from Egypt to the Indian Ocean it remained a fertile breeding-ground of nations.

These early civilisations are merely the highest point of Neolithic culture. The Egyptian remains show a very gradual development of pottery, ornamentation, etc., into which copper articles are introduced in time. The dawn of civilisation is as gradual as the dawn of the day. The whole gamut of culture—Eolithic, Palaeolithic, Neolithic, and civilised—is struck in the successive layers of Egyptian remains. But to give even a summary of its historical development is neither necessary nor possible here. The maintenance of its progress is as intelligible as its initial advance. Unlike China, it lay in the main region of human development, and we find that even before 6000 B.C. it developed a system of shipping and commerce which kept it in touch with other peoples over the entire region, and helped to promote development both in them and itself.

Equally intelligible is the development of civilisation in Mesopotamia. The long and fertile valley which lies between the mountainous region and the southern desert is, like the valley of the Nile, a quite recent formation. The rivers have gradually formed it with their deposit in the course of the last ten thousand years. As this rich soil became covered with vegetation, it attracted the mountaineers from the north. As I said, the earliest centre of the civilisation which was to culminate in Babylon and Nineveh is traced at Susa, on the hills to the north, about 6000 B.C. The Akkadians (highlanders) or Sumerians, the Turanian people who established this civilisation, descended upon the rivers, and, about 5000 B.C., set up the early cities of Mesopotamia. As in the case of Egypt, again, more tribes were attracted to the fertile region, and by about 4000 B.C. we find that Semitic tribes from the north have superseded the Sumerians, and taken over their civilisation.

In these ancient civilisations, developing in touch with each other, and surrounded by great numbers of peoples at the high Neolithic level from which they had themselves started, culture advanced rapidly. Not only science, art, literature, commerce, law, and social forms were developed, but moral idealism reached a height that compares well even with that of modern times. The recovery in our time of the actual remains of Egypt and Babylon has corrected much of the libellous legend, which found its way into Greek and

371

European literature, concerning those ancient civilisations. But, as culture advances, human development becomes so complex that we must refrain from attempting to pursue, even in summary, its many outgrowths. The evolution of morality, of art, of religion, of polity, and of literature would each require a whole volume for satisfactory treatment. All that we can do here is to show how the modern world and its progressive culture are related to these ancient empires.

The aphorism that "all light comes from the east" may at times be pressed too literally. To suggest that western peoples have done no more than receive and develop the culture of the older east would be at once unscientific and unhistorical. By the close of the Neolithic age a great number of peoples had reached the threshold of civilisation, and it would be extremely improbable that in only two parts of the world the conditions would be found of further progress. That the culture of these older empires has enriched Europe and had a great share in its civilisation, is one of the most obvious of historical truths. But we must not seek to confine the action of later peoples to a mere borrowing of arts or institutions.

Yet some recent historical writers, in their eagerness to set up indigenous civilisations apart from those of Egypt and Mesopotamia, pass to the opposite extreme. We are prepared to find civilisation developing wherever the situation of a people exposes it to sufficient stimulation, and we do find advance made among many peoples apart from contact with the great southern empires. It is uncertain whether the use of bronze is due first to the southern nations or to some

European people, but the invention of iron weapons is most probably due to European initiative. Again, it is now not believed that the alphabets of Europe are derived from the hieroglyphics of Egypt, though it is an open question whether they were not derived, through Phoenicia, from certain signs which we find on ancient Egyptian pottery.

If we take first a broad view of the later course of civilisation we see at a glance the general relation of east and west. Some difficulty would arise, if we pressed, as to the exact stage in which a nation may be said to become "civilised," but we may follow the general usage of archaeologists and historians. They tell us, then, that civilisation first appears in Egypt about 8000 B.C. (settled civilisation about 6000 B.C.), and in the Mesopotamian region about 6000 B.C. We next find Neolithic culture passing into what may be called civilisation in Crete and the neighbouring islands some time between 4000 and 3000 B.C., or two thousand years after the development of Egyptian commerce in that region. We cannot say whether this civilisation in the AEgean sea preceded others which we afterwards find on the Asiatic mainland. The beginning of the Hittite Empire in Asia Minor, and of Phoenician culture, is as yet unknown. But we can say that there was as yet no civilisation in Europe. It is not until after 1600 that civilisation is established in Greece (Mycenae and Tiryns) as an offshoot of AEgean culture. Later still it appears among the Etruscans of Italy—to which, as we know, both Egyptian and AEgean vessels sailed. In other words, the course of civilisation is very plainly from east to west.

But we must be careful not to imagine that this represents a mere transplantation of southern culture on a rude northern stock. The whole region to the east of the Mediterranean was just as fitted to develop a civilisation as the valley of the Nile. It swarmed with peoples having the latest Neolithic culture, and, as they advanced, and developed navigation, the territory of many of them became the high road of more advanced peoples. A glance at the map will show that the easiest line of expansion for a growing people was westward. The ocean lay to the right of the Babylonians, and the country north and south was not inviting. The calmer Mediterranean with its fertile shores was the appointed field of expansion. The land route from Egypt lay, not to the dreary west in Africa, but along the eastern shore of the Mediterranean, through Syria and Asia Minor. The land route from Babylon lay across northern Syria and Asia Minor. The sea route had Crete for its first and most conspicuous station. Hence the gradual appearance of civilisation in Phoenicia, Cappadocia, Lydia, and the Greek islands is a normal and natural outcome of the geographical conditions.

But we must dismiss the later Asiatic civilisations, whose remains are fast coming to light, very briefly. Phoenicia probably had less part in the general advance than was formerly supposed. Now that we have discovered a powerful civilisation in the Greek islands themselves, we see that it would keep Tyre and Sidon in check until it fell into decay about 1000 B.C. After that date, for a few centuries, Phoenicia had a great influence on the development of Europe. The Hittites, on the other hand, are as yet imperfectly known. Their main region was Cappadocia, where, at least as far back as 1500 B.C., they developed so characteristic a civilisation, that its documents or inscriptions are almost undecipherable. They at one time overran the whole of Asia Minor. Other peoples such as the Elamites, represent similar offshoots of the fermenting culture of the region. The Hebrews were probably a small and unimportant group, settled close round Jerusalem, until a few centuries before the Christian Era. They then assimilated the culture of the more powerful nations which crossed and recrossed their territory. The Persians were, as we saw, a branch of the Aryan family which slowly advanced between 1500 and 700 B.C., and then inherited the empire of dying Babylon.

The most interesting, and one of the most recently discovered, of these older civilisations, was the AEgean. Its chief centre was Crete, but it spread over many of the neighbouring islands. Its art and its script are so distinctive that we must recognise it as a native development, not a transplantation of Egyptian culture. Its ruins show it gradually emerging from the Neolithic stage about 4000 B.C., when Egyptian commerce was well developed in its seas. Somewhere about 2500 B.C. the whole of the islands seem to have been brought under the Cretan monarchy, and the concentration of wealth and power led to a remarkable artistic development, on native lines. We find in Crete the remains of splendid palaces, with advanced sanitary systems and a great luxuriance of ornamentation. It was this civilisation which founded the centre at Mycenae, on the Greek mainland, about the middle of the second millennium B.C.

But our inquiry into the origin of European civilisation does not demand any extensive description of the AEgean culture and its Mycenaean offshoot. It was utterly destroyed between 1500 and 1000 B.C., and this was probably done by the Aryan ancestors of the later Greeks or Hellenes. About the time when one branch of the Aryans was descending upon India and another preparing to rival decaying Babylonia, the third branch overran Europe. It seems to have been a branch of these that swept down the Greek peninsula, and crossed the sea to sack and destroy the centres of AEgean culture. Another branch poured down the Italian peninsula; another settled in the region of the Baltic, and would prove the source of the Germanic nations; another, the Celtic, advanced to the west of Europe. The mingling of this semi-barbaric population with the earlier inhabitants provided the material of the nations of modern Europe. Our last page in the story of the earth must be a short account of its civilisation.

The first branch to become civilised, and to carry culture to a greater height than the older nations had ever done, was the Hellenes. There is no need for us to speculate on the "genius" of the Hellenes, or even to enlarge on the natural advantages of the lower part of the peninsula which they occupied. A glance at the map will explain why European civilisation began in Greece. The Hellenes had penetrated the region in which there was constant contact with all the varied cultures of the older world. Although they destroyed the AEgean culture, they could not live amidst its ruins without receiving some influence. Then the traders of Phoenicia, triumphing in the fall of their AEgean rivals, brought the great pacific cultural influence of commerce to bear on them. After some hundreds of years of internal trouble, barbaric quarrels, and fresh arrivals from the north, Greece began to wear an aspect of civilisation. Many of the Greeks passed to Asia Minor, as they increased, and, freed from the despotism of tradition, in living contact with the luxury and culture of Persia, which had advanced as far as Europe, they evolved the fine civilisation of the Greek colonies, and reacted on the motherland. Finally, there came the heroic struggle against the Persian invaders, and from the ashes of their early civilisation arose the marble city which will never die in the memory of Europe.

The Romans had meantime been advancing. We may neglect the older Italian culture, as it had far less to do with the making of Italy and Europe than the influence of the east. By about 500 B.C. Rome was a small kingdom with a primitive civilisation, busy in subduing the neighbouring tribes who threatened its security, and unconsciously gathering the seeds of culture which some of them contained. By about 300 B.C. the vigour of the Romans had united all the tribes of Italy in a powerful republic, and wealth began to accumulate at Rome. Not far to the east was the glittering civilisation of Greece; to the south was Carthage, a busy centre of commerce, navigation, and art; and from the Mediterranean came processions of ships bringing stimulating fragments and stories of the hoary culture of the east. Within another two hundred years Rome annihilated Carthage, paralysed and overran Greece, and sent its legions over the Asiatic provinces of the older empires. By the beginning of the Christian Era all that remained of the culture of the old world was gathered in Rome. All the philosophies of Greece, all the religions of Persia and Judea and Egypt, all the luxuries and vices of the east, found a home in it. Every stream of culture that had started from the later and higher Neolithic age had ended in Rome.

And in the meantime Rome had begun to disseminate its heritage over Europe. Its legions poured over Spain and Gaul and Germany and Britain. Its administrators and judges and teachers followed the eagles, and set up schools and law-courts and theatres and baths and temples. It flung broad roads to the north of Britain and the banks of the Rhine and Danube. Under the shelter of the "Roman Peace" the peoples of Europe could spare men from the plough and the sword for the cultivation of art and letters. The civilisations of Britain, France, Germany, Spain, North Africa, and Italy were ushered into the calendar of mankind, and were ready to bear the burden when the mighty city on the Tiber let the sceptre fall from its enfeebled hands.

Rome fell. The more accurate historians of our time correct the old legend of death from senile decay or from the effect of dissipation. Races of men, like races of animals, do not die; they are killed. The physical deterioration of the citizens of Rome was a small matter in its fall. Fiscal and imperial blunders loosed the frame of its empire. The resources were still there, but there was none to organise and unify them. The imperial system—or chaos—ruined Rome. And just when the demoralisation was greatest, and the Teutonic tribes at the frontiers were most numerous and powerful, an accident shook the system. A fierce and numerous people from Asia, the Huns, wandered into Europe, threw themselves on the Teutonic tribes, and precipitated these tribes upon the Empire. A Diocletian might still have saved the Empire, but there was none to guide it. The northern barbarians trod its civilisation underfoot, and Europe passed into the Dark Ages.

One more application of the evolutionary principle, and we close the story. The "barbarians"—the Goths and Vandals and their Germanic cousins—were barbaric only in comparison with the art and letters of Rome. They had law, polity, and ideals. European civilisation owes elements to them, as well as to Rome. To say simply that the barbarians destroyed the institutions of Rome is no adequate explanation of the Dark Ages. Let us see rather how the Dark Ages were enlightened.

It is now fully recognised that the reawakening of Europe in the twelfth and thirteenth centuries was very largely due to a fresh culture-contact with the older civilisations. The Arabs had, on becoming civilised, learned from the Nestorians, who had been driven out of the Greek world for their heresies, the ancient culture of Greece. They enshrined it in a brilliant civilisation which it inspired them to establish. By the ninth century this civilisation was exhibited in Spain by its Moorish conquerors, and, as its splendour increased, it attracted the attention of Europe. Some

Christian scholars visited Spain, as time went on, but the Jews were the great intermediaries in disseminating its culture in Europe. There is now no question about the fact that the rebirth of positive learning, especially of science, in Europe was very largely due to the literature of the Moors, and their luxury and splendour gave an impulse to European art. Europe entered upon the remarkable intellectual period known as Scholasticism. Besides this stimulus, it must be remembered, the scholars of Europe had at least a certain number of old Latin writers whose works had survived the general wreck of culture.

In the fifteenth century the awakening of Europe was completed. The Turks took Constantinople, and drove large numbers of Greek scholars to Italy. Out of this catastrophe issued the great Renaissance, or rebirth, of art, science, and letters in Italy, and then in France, Germany, and England. In the new intellectual ferment there appeared the great artists, great thinkers and inventors, and great navigators who led the race to fresh heights. The invention of printing alone would almost have changed the face of Europe. But it was accompanied by a hundred other inventions and discoveries, by great liberating and stimulating movements like the Reformation, by the growth of free and wealthy cities, and by the extension of peace over larger areas, and the concentration of wealth and encouragement of art which the growth and settlement of the chief European powers involved. Europe entered upon the phase of evolution which we call modern times.

The future of humanity cannot be seen even darkly, as in a glass. No forecast that aspires beyond the immediate future is worth considering seriously. If it be a forecast of material progress, it is rendered worthless by the obvious consideration that if we knew what the future will do, we would do it ourselves. If it is a forecast of intellectual and social evolution, it is inevitably coloured by the intellectual or social convictions of the prophet. I therefore abstain wholly from carrying the story of evolution beyond realities. But I would add two general considerations which may enable a reflective reader to answer certain questions that will arise in his mind at the close of this survey of the story of evolution.

Are we evolving to-day? Is man the last word of evolution? These are amongst the commonest questions put to me. Whether man is or is not the last word of evolution is merely a verbal quibble. Now that language is invented, and things have names, one may say that the name "man" will cling to the highest and most progressive animal on earth, no matter how much he may rise above the man of to-day. But if the question is whether he WILL rise far above the civilisation of to-day, we can, in my opinion, give a confident answer. There is no law of evolution, but there is a fact of evolution. Ten million years ago the highest animal on the earth was a reptile, or, at the most, a low, rat-like marsupial. The authorities tell us that, unless some cosmic accident intervene, the earth will remain habitable by man for at least ten million years. It is safe to conclude that the man of that remote age will be lifted above the man of to-day as much as we transcend the reptile in intelligence and emotion. It is most probable that this is a quite inadequate expression of the future advance. We are not only evolving, but evolving more rapidly than living thing ever did before. The pace increases every century. A calm and critical review of our development inspires a conviction that a few centuries will bring about the realisation of the highest dream that ever haunted the mind of the prophet. What splendours lie beyond that, the most soaring imagination cannot have the dimmest perception.

And the last word must meet an anxiety that arises out of this very confidence. Darwin was right. It is— not exclusively, but mainly—the struggle for life that has begotten higher types. Must every step of future progress be won by fresh and sustained struggle? At least we may say that the notion that progress in the future depends, as in the past, upon the pitting of flesh against flesh, and tooth against tooth, is a deplorable illusion. Such physical struggle is indeed necessary to evolve and maintain a type fit for the struggle. But a new thing has come into the story of the earth— wisdom and fine emotion. The processes which begot animal types in the past may be superseded; perhaps must be superseded. The battle of the future lies between wit and wit, art and art, generosity and generosity; and a great struggle and rivalry may proceed that will carry the distinctive powers of man to undreamed-of heights, yet be wholly innocent of the passion-lit, blood-stained conflict that has hitherto been the instrument of progress.